The Manifesting Mind

REWIRE YOUR BRAIN

TO ENGINEER YOUR DREAM LIFE

The Manifesting Mind

REWIRE YOUR BRAIN
TO ENGINEER YOUR DREAM LIFE

STEPHANIE PIERUCCI

PIERUCCI
PUBLISHING
Elevating World Consciousness
Through Books

Publishing@PierucciPublishing.com

Visit our website at www.pieruccipublishing.com

Paperback ISBN: 978-0-578839-60-8
Ebook ISBN: 978-1-956257-73-1
Hardcover ISBN: 978-1-956257-74-8

Library of Congress Control Number: 2023902772

Cover Design by Stephanie Pierucci

"Ask, and it shall be given to you;
Seek, and you shall find;
Knock, and it shall be opened unto you."
Matthew 7:7, RSV

Dedication

For Hunter. I'm so happy you were born. Thank you for expanding my heart beyond anything I thought I could feel. Thank you for giving me the honor of being your mom and teaching me more every day.

Dad, thank you for teaching me to run fast, to stop whining, to never give up, and for entertaining me relentlessly with your side-splitting wit and cleverness. I'm empowered by your faith in me.

Mom, thank you for showing me a love for books; your nose is always buried in one! Thank you for turning a blind eye on my book addiction as a kid, even though it must have infuriated you to find me reading until 4 a.m. on school nights. I promise it's paying off.

Papa, thank you for teaching me about entrepreneurship over many late nights in your house. You illustrated to me the generous dividends of belief in oneself, taking calculated risks, and having integrity in life and business.

Nanny, thank you for molding me with your artistic gift and making me your protégé over countless hours of drawing, painting, and sketching. Thank you for teaching me that you can be aware of evil in the world while still celebrating the light and embodying love every day.

Grandpa Jerry, thank you for modeling unconditional love, compassion, non-judgment, equanimity, and insatiable love for earth and nature. I imagine I've chosen to live in Colorado in part because of my childhood spent indulging in exploring cool rocks with you.

Grandma, thank you for infusing me with a love for God and hunger to make my life a blessing. Among the most important gifts you've given me is the steady channel of joyful songs that is never far from my conscious mind. I know that these songs have lifted me out of some dark spots and times; perhaps more times than I can conceive.

Bel, thank you for showing me that I am worthy of a love chosen from a healed, empowered place and proving to me that I have broken so many of the curses and cords I came here with. I pray that our children's children will now be free of narcissistic abuse because of the healing we've done together. Thank you for living a life of yoga from sunrise to sunset and all the powerful manifestation moments in between.

Acknowledgments

To Lauryn Maloney-Gepfert. Thank you for inspiring this work and helping me remember how beautifully and wonderfully we are made. When I met you, the implicit goal was to assist the Neurofunctional Institute with some local marketing. I arrived beaten, bruised, and battered after a harrowing divorce and several years of single motherhood having taken its toll on my spirit and body. You chose to invest our time together that day helping me learn to heal my body and reclaim my birthright as a daughter of one Almighty, Omnipotent God. I hope this book serves to illustrate the power of your teaching and the One who teaches through you.

To Jim Poole. Thank you for investing your life into making NuCalm an accessible tool to hundreds of thousands of people to heal their trauma from the inside out. When I learned about NuCalm, I was already many years into my work with neuroplasticity, but I had become stagnant in my power due to a block in my brain where trauma hadn't fully healed. The first time I used NuCalm, I slept for nineteen hours straight. Some months later I said, "Jim, I used to always fall asleep whenever I listened to NuCalm, but now I stay awake in a lucid dreamlike state throughout the entire recording." You laughed and said, "Well, Steph, that means it's working. You have healed so much trauma that your brain doesn't shut off every time you listen to the program. You're awake... in more ways than one!" I pray that one million more souls will experience even a fraction of the healing I've enjoyed, so much of which I cannot yet consciously comprehend.

To Jamie Prickett. Thank you for being an inspiration in the power of manifesting abundance and prosperity for your family and the families of countless members of the Experior Financial Group team. Writing your book with you added rocket fuel to my own ability to manifest wealth in my life. Your life embodies a marriage between grit and grace. Above all, you illustrated to me that with honest hard work toward a worthy goal, anything can be achieved.

To Dr. Heather Gessling. Thank you for being the very embodiment of the mama bear. Any man or woman would be blessed to have your success as a respected doctor in a private practice, but you are also among the world's leading medical freedom fighters at the tip of the spear in forging a parallel medical system. You risked losing everything so that you could stand in truth, and you have done it with kindness, gentleness, self-control, and every fruit of your loving spirit. It's not often that we get to watch leaders who have achieved their success with integrity, and I pray that the whole world will one day read your book and model their lives after you. It is possible to change the world without playing at the world's level.

To Dr. Peter A. McCullough. Thank you for paving the way on the legal and medical fronts to lead humanity closer to truth. Your integrity in medicine and your tireless determination to be a light in a dark world fuels me to do good work without ceasing, to run without growing weary, and to remember that I can do all things through Christ who gives me strength. One day I asked you whether or not you believed in the devil. To this day, I'm not sure what compelled me to ask such a question! You replied that although you were always of the conviction that there is a God, seeing the evil in the world today, you now know that there is an enemy; there is a devil. I'm proud to stand alongside you in the resistance to that devil's lies, distortions, and schemes.

Publisher's Note

PLEASE READ

This book is meant to narrate one woman's subjective journey while learning about how to use the power of her mind for a happy and healthy life. The information provided by the author, website, or this company is not a substitute for a face-to-face consultation with your healthcare providers and should not be construed as individual medical advice. If a condition persists, please contact your healthcare providers. This book is provided for personal and informational purposes only. This book is not to be construed as any attempt to either prescribe or practice medicine. Neither is the site to be understood as putting forth any cure for any type of acute or chronic health problem. You should always consult with a competent, fully licensed healthcare professional when making any decision regarding your health. The owners of this site will use reasonable efforts to include up-to-date and accurate information on this internet site, but make no representations, warranties, or assurances as to the accuracy, currency, or completeness of the information provided. The owners of this site shall not be liable for any damages or injury resulting from your access to, or inability to access, this Internet site, or from your reliance upon any information provided on this site. All rights reserved. No part of this publication may be reproduced, transmitted, transcribed, stored in a retrieval system, or translated into any language, in any form, by any means, without the written permission of the author.

Table of Contents

Dedication   *vii*

Acknowledgments   *ix*

Publisher's Note   *xi*

Table of Contents   *xiii*

A Note About the Second Edition of The Manifesting Mind   1

Cliff Notes   8

Prologue   9

Introduction   11

PART 1

Chapter One: Learning to Manifest   23

Chapter Two: What is Consciousness?   37

Chapter Three: The Law of Nature and the Yin-Yang   47

Chapter Four: Turning Sludge into Sugar   55

Chapter Five: The Brain Experts Who Saved My Life   59

Chapter Six: The Science of Manifestation   63

Chapter Seven: Do We "Create" The Bad Things in Our Lives?   67

Chapter Eight: What is Disease?   71

Chapter Nine: Next Level Brain Healing with NuCalm   77

Chapter Ten: Our Thoughts Create Our Realities   83

Chapter Eleven: Manifestation is Possible for Anybody   87

Chapter Twelve: Where Spirit Meets Science   91

Chapter Thirteen: Interrupting Subconscious Patterns to Manifest   101

Chapter Fourteen: The Power of Relaxation   105

Chapter Fifteen: The Science of Happiness   107

Chapter Sixteen: Smiling and Manifestation   117

Chapter Seventeen: Manifesting Money   121

PART 2
MANIFESTATION STORIES

Chapter Eighteen: How Manifestation Looks In "Real Life"   131

Chapter Nineteen: Mountain Lions and Letting Go   135

Chapter Twenty: Manifesting Safety from a Child Predator   141

Chapter Twenty-One: Manifesting A Child   147

Chapter Twenty-Two: Manifesting My Perfect Home   151

Chapter Twenty-Three: Manifesting $25,000 In A Week   155

Chapter Twenty-Four: Manifesting My Perfect Lover   157

Chapter Twenty-Five: Manifesting Healing After Infidelity   159

Chapter Twenty-Six: Manifesting Healing After Divorce   163

Chapter Twenty-Seven: Manifesting Epic Love After Divorce   167

Chapter Twenty-Eight: Manifesting My Dream Job   169

PART 3

7 Steps to Start Manifesting Today   175

Step One: Reprogramming Thoughts with Mind-Rewiring Language   177

Step Two: Leaning into Manifestation with Intentional Relaxation   183

Step Three: Meditation   193

Step Four: Creating Intentions   203

Step Five: Breathwork for Manifestation   209

Step Six: Manifest Through Movement   215

Step Seven: Surrender   221

We Want to Hear from You!   228

Closing   229

Words to Know   231

Resources for Further Study   236

Books   237

Lectures/ Webinars/ Talks   246

Apps   248

"I discovered that when I believed my thoughts, I suffered, but that when I didn't believe them, I didn't suffer, and that this is true for every human being. Freedom is as simple as that. I found that suffering is optional. I found a joy within me that has never disappeared, not for a single moment."

BYRON KATIE

A NOTE ABOUT
THE SECOND EDITION OF
The Manifesting Mind

During a 2022 recommitment to my faith in Jesus Christ, I wondered many times if this manuscript would have to change, in turn, to align with my reinvigorated convictions and beliefs. Happily, this manuscript was originally written with sufficient soundness in science and without common Spirituality and New Age gimmicks and/or superstitions that the manuscript stands almost untouched in this second version.

The book is rooted in empirical evidence and defensible philosophies available to any audience who wants to understand how to harness the power of their mind. Like scientists and teachers Dr. Caroline Leaf, Dr. David Hawkins, and Lauryn Maloney-Gepfert, I use the word God when delivering a message that is nevertheless accessible to people who prefer the word Spirit, Source, Universe, Goddess, Buddha, or any other name you assign to a creative or guiding force. My book talks about meditation, which many Christians decide to swap out with the word prayer. In fact, many Christians have previously and will continue to say that this book is a work of heresy by the very use of the word *Manifestation* due to its association with more New Age or Spirituality movements.

In this book, the word *manifest* or *manifestation* is used to represent, in gross and broadly sweeping terms, "the mindset used to materialize our thoughts into reality." Dr. David R. Hawkins definition is also outstanding: **"The emergence of potentiality out of essence into manifestation."**

When I use the word manifestation, rather than conjuring up images of the evil eye, tarot cards, crystal balls, or other otherwise occult practices, I would ask you to see these terms: *emergence, freedom, and power.*

For me, freedom involves grace. I see the image of a dove when I think of the word manifestation. I am free from limiting beliefs. I am free from trauma that manifests into disease or discomfort in my body, and I am free to create a life of wealth, abundance, and prosperity, because I so choose.

My power to manifest is my ability to be unchained from the persistent messages in my past, present, and inevitable future which tether me to trauma.

An Update on My Life Since the First Edition

I don't presume the average reader picking up this book cares too much about me, the author, but rather about the science that will lead them to a freer, more powerful life. In that case, please skip to the Introduction.

However, since this book was published alongside a 5-day Manifestation event with participants with whom I keep in touch; many of whom I've coached over the years, those readers will, in fact, care about the evolution I've experienced since first publishing.

Aside from the particularly critical eye on the manuscript with regards to my Christian faith, I have chosen to release this second edition of *The Manifesting Mind* with an intention to celebrate some of my own "manifestations" since the first edition.

When I first published this book, I shared details about both my victories as well as my struggles in life at that very moment in time. I promised the Reader that I would update them with regards to the struggles that I would manifest into victories. The manuscript will update the Reader on those victories so that they can use some of the tools I present in this manuscript to produce a greater effect in their own lives. Some of the previous storylines have now achieved resolution, and I have the honor of offering a greater perspective on those stories.

Turning Lemons into Lemonade

In 2020, when this book was being written, I was going through an awakening while many others were in their dark night of the soul; manipulated by fear porn and propaganda from the planned-demic, manipulating them to unnecessarily isolate from loved ones, frighten and abuse their children with masks, and surrender freedoms to a nefarious global elite. In short, I saw the Covid-19 situation for what it was—a tool by corrupt, wealthy actors to enslave the common person psychologically and physically. After an exhausting 2020 making little impact in my local Aspen community, shouting from the rooftops that masks were unnecessary, if not dangerous, and that the coming vaccine would invariably injure and even murder countless people worldwide, I finally gave up. I surrendered, but in a defeated, disempowered way.

I felt angry, betrayed, isolated, and bewildered that my opinion, which is now held as a fact by an increasing majority of my global peers, seemed insane. *I* seemed insane to my local Aspen community. Battle wearied and seemingly defeated, I gave up on trying to be a teacher or leader to others. I stopped running my Manifesting Secrets training, stopped emailing our digital newsletter following, stopped posting on our Facebook and Instagram pages, and kept my head down building Pierucci Publishing. In some ways, with most of my fellow Americans still under the deception that the Covid-19 bioweapon was worth locking down, masking, and even receiving experimental gene-based therapy for, I somewhat gave up on humanity.

During that time, I was also subject to relentless and horrific legal and financial abuse by a former partner who was dead set on hurting me by trying to force the courts to mandate the Covid-19 experimental gene-based therapy called a "vaccine" on my only child, my son Hunter. As it so often goes, in that dark night of the soul, I called out to the Creator for help.

Although some people encounter God with a flash of lightning or a spontaneous healing, my revelation was slower, although no less powerful and blindingly clear.

That Dark Night Illustrated a Lost Prophecy

When I was eighteen years old, a traveling prophet, the first and only one I've ever known to date, came through my town and gathered a group of Christians in a friend's home. We all sat in a pastel, eighties-decorated living room on peach-colored carpets that were dated even then. The prophet pointed to me and said, "I have a message for you from God, young lady. Do I have your permission to share this vision?"

After I gave the prophet my permission to illustrate what she saw, she told me that she saw me at a desk writing furiously. As I wrote, books piled up in front of me. Soon, there were piles of books all over the desk. Suddenly I stopped writing and grabbed a book from the pile. That book then lifted me and carried me; I was soaring with this book as my jetpack. That book, of course, was the Bible.

For some years, I resented this vision when I was working in finance, hating every second of my existence. I thought I'd die sitting at a desk crunching numbers, trying to sell people financial instruments I didn't believe in.

One day after drinking an entire bottle of wine by myself, I began ("drunk?") dialing local publishing companies looking for a job. I left messages telling the CEOs I'd do anything for the chance to work around books. And I meant it.

Within weeks, I started working with a publishing company in Chicago, and over the past thirteen years I've done little else but create content with words. Even after becoming a single mother, I wrote and published a book on baby travel, created a course and book on Confidence, created the massive Manifesting Secrets Brain Training program, and a vault of resources for mothers called "Moms Wear Capes" that held over sixty hours of video content. Although I wasn't working "full time" when my son was small, I couldn't stop creating books and courses.

It was my destiny.

In the first two years of starting my current publishing company, Pierucci Publishing, I consulted over 250 authors and published scores of books. I was fulfilling the vision that had been prophesied over me... *almost.*

In late 2021, my arduous and financially straining legal battle came to an end with an abrupt settlement. The highest tax, undoubtedly, was the commitment the legal battle required of my time and energy. I had finally worn down the opposing party with my refusal to give up, and the experimental gene-based therapy called a "vaccine" had started to injure and kill countless young people. The opposing party's case to inject my son was going to be lost soon, and with that, their credibility with a judge. They undoubtedly want to appear before again in court someday for another excessive, frivolous, and abusive legal motion or suit.

Once my free time wasn't filled with requests from my lawyer, preparation for trial, enlisting witnesses, tidying up my evidence and exhibits, and earning the money to pay my high-priced attorneys, I had a moment to sit back and reflect on what had happened over the past few years with sobriety, clarity, and perspective.

Before my eyes I saw a timeline of events that led me to founding Pierucci Publishing about the time when I published this book. In 2021 and 2022, I had the honor of working with influential thought leaders through my publishing company, particularly those in the medical freedom fighting movement. Last year alone, I had the honor of writing two Wall Street Journal Bestselling books and I'm even now becoming involved with screenplays, motion pictures, and documentaries. My work takes me around the country sitting in the living rooms and offices of some of the most influential people in the world, mapping out our collective path to a freer society for all of humanity. I'm proud to be at the forefront (although still behind the scenes) of such a movement.

Best of all, because of Pierucci Publishing, I was continually introduced to the precise parties who would help fuel my legal battle to protect Hunter. My personal life fueled my work, and vice versa. It was stunning to behold the synergy between the two worlds. The word *synchronicity* began to have a new meaning.

For this and so many other reasons, I saw Pierucci Publishing as a direct result of my manifestation power, but I also saw that Pierucci Publishing, a company devoted to elevating world consciousness through books, was *the fulfillment of what that prophet spoke over me nearly twenty-two years ago.*

Above all, I saw how the authors God sent my way were sent to remind me of my purpose as a channel of God's truth.

What if that prophet didn't have good intentions? What if I gave her words total power and control over my life, like so many do when seeking fortune tellers, mediums, or tarot card readers?

To this day, I can't tell you whether there is a Biblical defense of that woman's practice. Your words are magic, they say, that's why they call it "spelling". This book will convince you, if nothing else, that words are spells. They manifest your reality. That woman's prophecy could have been a blessing... or a curse. I choose to hold it as the former.

Language Notes

When I first wrote this book, I used the words God and Universe interchangeably at times. That's no longer the case.

But even more importantly, there's been a shift in my thinking away from the concept that we are gods or goddesses. In fact, I wholly refute the concept that we are gods and goddesses, finding that to be a dangerous, narcissistic notion designed to rob the One True God of His power.

When I first wrote this book, I was knee deep in the Spirituality movement and generally negative toward religion. I have dramatically changed my thinking. I find that the Spirituality movement is rooted in ego and that the insistence that religion is wrong and Spirituality is right has led many people to an isolated, lonely life. What's more, seeing the exodus from our churches is destructive to the family, to our communities, and to the inner fortitude gained from faithfully showing up, even if merely on Sundays, for a ceremony or celebration designed to help us put everything else away and focus on God. I don't want to be my own spiritual boss anymore. In my eyes, there's one boss and there is a 100% chance that prosperity and eternal happiness will happen if I hand Him the wheel.

In my experience, much of New Age spirituality is designed to bring people further away as opposed to closer to their true source of power. Tarot cards, mediums, seers, and other gimmicks put other human beings or, in the case of tarot and angel cards, cheap pieces of paper, between an individual and his or her relationship with God.

Although the first edition of this book was a gentle account from a "Christian Mystic," this second version has been slightly edited to:

A) Account for my recommitment to my faith.
B) Update the reader on my successful legal battle as well as the success of my growing company, which were two manifestations from the first book I promised to update the reader about in a later version.
C) Weave more stories of *The Manifesting Mind* tools I use in my remarkable life into this second edition so that you will be further empowered and encouraged by a reminder of how beautifully, wonderfully, and fearfully you were made.

There is one final reason I have felt the call to publish this second edition of *The Manifesting Mind*. We have collectively been traumatized by events succeeding a time roughly around March of 2020. Although I've taken a resolute stance on my beliefs with regards to the matter, my heart is compassionately broken for those who've lost family members and friends to the Covid-19 bioweapon or subsequent experimental "vaccine". In response to that collective pain, this book is more important than ever. It is for that reason that I have included mention of my journey with NuCalm as a trauma recovery tool that both my son and I use regularly. Together, as we heal ourselves, we will create a more beautiful world for my child and yours.

It is from your most healed, most empowered state that you will be the most useful to the future of humanity. And more than ever, we need you to be in your very best fighting form for the days ahead.

Blessings to you in your healing journey,

Stephanie Pierucci
February, 2023

Cliff Notes

Just by picking up this book and opening yourself up to receiving the wisdom of manifestation, you have already begun to manifest. When we do something new, it interrupts the grooves and pathways that the synapses in our brains normally follow. It opens us up to neuroplasticity, rewiring the mind through mental intention. In a later chapter we'll talk about the two fundamental principles of neuroplasticity, but I'll attempt to repeat them a few times throughout this book so that they begin to become something of which you're consciously, effortlessly aware.

The first is this: **neurons that fire together, wire together**. What we do, think, or say repeatedly becomes a subconscious habit; it becomes part of the neurological structure of our brains. This means that we can manifest good things or bad things; it depends on the habits you create.

The second principle is: **what you don't use, you lose**.

In short, healing the brain from negative thoughts that beget a negative life is a matter of intention and consistent practice. This book will detail how intention materializes healing, abundance, and health as well as practices that will help you the reader develop the manifestation muscles to create a more positive mindset and, therefore, life.

Prologue

Everything you see around you was once a thought. The chair you're sitting in. The water filter on your kitchen counter. Your partner and the relationship you share. Your bank account balance. All these things have been manifested; brought into existence as the results of thoughts; either positive or negative.

In this book, you'll go on a journey to understand how conscious and subconscious thoughts have brought forth prosperity or poverty in different aspects of your life: primarily focusing on love, health, wealth, and happiness.

In Part One of this book, you'll learn the science of thought and how it works on a philosophical level. Then we'll bring the plane down and dive into the mechanics of thoughts. Thoughts have physical structure in our brains; they're the results of synapses that have been programmed over time, usually decades.

In Part Two of this book, you'll learn the practical ways that thoughts have created different outcomes in my life; and how tweaks in my own thoughts, which I refer to as "mental-chiropractics" or micro-adjustments, have either begat persistent, repeated pain and suffering, or have provided almost miraculous healing and relief in my life and body.

Due to my work with many doctors, healers, and scientists, I've included many stories about healing the body through neuroplasticity; the science of materializing our mental intentions. However, although I spend some time discussing the healing capacity of manifestation, I am not a doctor, scientist, nor do I consider myself a healer. Your mission is to hear these stories as inspiration, but please, always talk to your healthcare provider about any physical discomfort, disease, or malady in conjunction with any of the inspirational stories and neuroplasticity practices I discuss in this book.

In Part Three of this book, you'll learn Seven Steps You Can Use to Start Manifesting Today. You don't need to consult a shaman in Peru and drink vomit-inducing teas to have a revelatory experience in your mind from the comfort of your home. In fact, I believe that one of the many ills of the modern Self-Help, Personal Development, Spirituality, and New Age movements is to separate you from your healing by making you think you need a yoga retreat or Ayahuasca ritual to see God. That's hogwash. The power is within you to remarkably heal your life and body right now.

If you are not a believer in God, the tools in this book will still apply to you. Whether or not you believe in a Divine Creator, you have been crafted in a magnificent, wonderful way. Your body is a healing machine. Your mind can transform itself. You are powerful beyond measure. If you don't believe anybody is listening when you pray, do it for yourself.

Listen to yourself.
Take this seriously, but have fun with it!

I'm honored to be on this healing journey with you.

Introduction

*"The amount that she loved us was beyond her reach.
It could not be quantified or contained. It was the ten
thousand things in the Tao Te Ching's Universe, and then
ten thousand more. It was full-throated and unadorned.
Each day she blew through her entire reserve."*
CHERYL STRAYED

The toggle between anxiety and depression, feelings of failure, and emotional distress and dysregulation were commonplace for most of the first thirty years of my life. Until I became a mother, most of my life was plagued with dark mental afflictions, suffering, and self-harm. After a nice boy I dated in high school left me for the perfect potential pastor's wife, my façade of being a righteous woman broke, and I spent more than a decade rebuilding the foundation of my faith until 2022. At that time I "got off the fence," so to speak, surrendering my life to the God who always was and always will be the guiding force in my life.

Regardless of your religious or spiritual beliefs, this book is for you. It's consciously and carefully rooted in objective practices that enhance but do not require faith in God.

In some ways, this book is a diary of how I made manifestation a consistent habit, gradually rewiring my mind for loving relationships in place of toxic ones; self-control and optimization of my body, personal healing capabilities; a blossoming and surprisingly successful career where I once sabotaged innumerable opportunities and connections around me; and sincere happiness in my life irrespective of events taking place at any given time.

Today I am living in the very world I most feared and crushing it; absolutely loving my life. I am fueled with passion and purpose in every moment of the day. I absolutely love my son and our life in Colorado.

When I was a young girl, I was terrified at the thought of becoming a single mother. At best it looked unglamorous and lonely, but at worst it looked like a curse or affliction that haunted me. I would think, "Yikes! How could any woman screw up that badly? How could you deprive a child of the most important thing in the world; a family, stability, and a two-parent household?" I was judgmental and lacked compassion; carrying with me so many indoctrinated messages that a woman could never survive on her own; that nobody worthwhile could want a single mother; and that such a life was manifested from "bad behavior".

Growing up, I observed single mothers who were hairdressers, accountants, teachers, nurses and such. They'd return home at night to take care of their households, make the meals, vacuum the floors, help with homework, tell stories, finally get the kids to bed, and then retire to their balconies for a cigarette or two after a day of doing everything on their own. Then they'd clean up. Then they'd go to bed. Alone. It looked awful to me.

They would file taxes alone. They'd spend holidays as a third wheel. They sat at home on Saturday nights, alone. Their kids were latchkey kids; at once they seemed wild and ill-monitored, and also saddled with more responsibilities and worries than the children from what I understood to be stable homes; "two-parent" homes.

I thought to myself, *What kind of living hell is it to not have a family? How does a woman raise children on her own? Do you even have leftover money at the end of the month for a trip to Target? Heavens, that must be horrible...*

As you can imagine, in my blue-collar town, single moms didn't do very well. They had a sort of scarlet letter, at least in my eyes. In 2016 at the age of 33, my childhood nightmare became my personal reality.

As I had imagined, single motherhood was damn tough. For my son's first five years I was perpetually at the end of my rope with exhaustion, grief, guilt, and self-doubt. I was deeply engrossed in personal development work and therapy, but I wasn't breaking patterns and forming healthy new ones; I was merely putting Band-Aids on deep infections in my brain and psyche.

Life seems to continually beat you up when you are unhealed. In response to this perceived stream of "bad luck," there are two mindsets: either you can play victim, or you can take personal responsibility. Your ability to do the latter is step one in living a healthier, happier and more successful life.

Victim mentality looked like this for me:

"I lost my marriage, it's because he was a narcissist who took advantage of me with manipulative tools like mirroring, love bombing, and eventually isolation from my friends and family and a smear campaign when I tried to leave. Poor me!"

"I'm not successful because I paid my way through college, waitressing and bartending until the wee hours in the morning to get through it. I didn't have time to network or interview for internships at great companies. Poor me!"

"I was exploited and/or assaulted by several men over my childhood and early adult years. This caused trauma that led to, among other things, eating disorders and psychological distress. These eating disordered behaviors were a barrier to success in work and relationships. Poor me!"

Seeing these "struggles," it's easy to see how one could play the victim card. But I noticed that many of the successful personal development coaches or entrepreneurs I looked up to had it just as hard, if not harder, than I had. It was at that moment that I realized I wasn't a victim; I was living in a swamp of my own sins; my own manifestations.

I manifested relationship failure, poverty, and even eating disorders. Were there traumas that contributed to these material outcomes? Absolutely. But it was my responsibility to heal those traumas. And it's your responsibility to do the same.

I don't think this is easy. And that's why I've written this book.

Your First Manifestation Exercise

Before we move into the meat and marrow of the practices in this book that I've distilled from so many great scientists and teachers, let's start the book by getting one thing straight: you are wonderfully made and worthy of love and success. Period.

When I began to look at myself through God's eyes, I began to allow healing into my life because I knew I was worthy of more than I'd previously allowed myself to have. I chose self-love, but not in the sense of defending my behavior or elevating myself above others. I chose to see the God qualities in me that make me worthy of healing.

Your first exercise will be to take the negative things you've experienced and choose to see your God qualities in them. I'll start:

> I love the woman who has so much faith that she married a veritable stranger in search of a better life than she'd had.

> I love the woman who loves her son so much that even though single motherhood is sometimes exhausting, she wouldn't give it up for anything on earth.

> I love the persistence of a woman who never stops learning and seeking healing in her life. I love that woman whose heart is so compassionate that she can't help but to teach others what she's learned.

No matter where you are in life, you have been either consciously or subconsciously involved in the decisions that got you here. The purpose of this book is to help you take subconscious negative programming and consciously reprogram it in order to make better decisions and break free from trauma that has kept your brain on autopilot down a path you wouldn't consciously choose for yourself.

I'll share with you practical, realistic, and tangible ways that you can engage in rewiring your mind every day like a prayer that's merely a thought away.

Do We Manifest the Bad Things in Our Lives?

Now for the question everybody asks me. Do I manifest bad things? Did I create this cancer? Did I manifest this divorce? Did I choose this sick child? Did I ask to live in poverty? If my thoughts create my reality, did I choose devastation, loneliness, divorce, and even assault?

Here's what I do know. Manifestation is always happening. Every choice we make, down to the way we smile or what route we take to work, comes from manifestation. We are always manifesting. Still, it wasn't "fair" that these things were happening, right? There's no way I "wanted" these things? The divorce, the exhaustion, the assault... I wasn't asking for them, so how did I manifest them?

Bad things happen. We don't manifest every single event around us, although we can choose to manifest both good and bad things. I believe in the collective consciousness and connectedness of every soul on earth. And I also believe in tragedy. I believe that until we reunite with the Creator in Heaven (and I'm not sure I know exactly what that reunion looks like), we're living in a world that is prone to lawlessness. We're living in a man's world, not God's Heaven. God is obviously not the boss here.

I believe that mankind is capable of monstrous things; things I know about but don't want to think about. If it isn't already blatantly obvious, I'm a firm believer in Free Will. It is the express priority of my life to elevate world consciousness because I understand that a majority of the world vibrates in a very low consciousness state, inflicting damage on themselves and others. Nobody chooses that their child is born sick or that they experience sexual assault. You might say that we choose to have the child, or we choose to be in a bar late at night where predators lurk, but I release the thought of "I manifest everything" and I have simply decided that my answer to this question is, "we manifest our responses to everything". I'll discuss this more in later chapters.

Recently, my best friend sent me a video on Instagram where the author quoted a charming fictional story about a boy debating with his teacher. I long to attribute the original author; nay, I long to attribute this text to my own brilliance. This story is, to the best of my understanding, merely a charming parable.

A Parable About Evil...

Author Unknown

"Does evil exist?"

The university professor challenged his students with this question. Did God create everything that exists? A student bravely replied, "Yes, he did!"

"God created everything?" the professor asked.

"Yes sir", the student replied.

The professor answered, "If God created everything, then God created evil since evil exists, and according to the principle that our works define who we are then God is evil". The student became quiet before such an answer. The professor was quite pleased with himself and boasted to the students that he had proven once more that the Christian faith was a myth.

Another student raised his hand and said, "Can I ask you a question, professor?"

"Of course", replied the professor.

The student stood up and asked, "Professor, does cold exist?"

"What kind of question is this? Of course it exists. Have you never been cold?" The students snickered at the young man's question.

The young man replied, "In fact, sir, cold does not exist. According to the laws of physics, what we consider cold is in reality the absence of heat. Every body or object is susceptible to study when it has or transmits energy, and heat is what makes a body or matter have or transmit energy. Absolute zero (-460 degrees F) is the total absence of heat; all matter becomes inert and incapable of reaction at that temperature. Cold does not exist. We have created this word to describe how we feel if we have no heat."

The student continued, "Professor, does darkness exist?"

The professor responded, "Of course it does."

The student replied, "Once again you are wrong, sir, darkness does not exist either. Darkness is in reality the absence of light. Light we can study, but not darkness. In fact we can use Newton's prism to break white light into many colors and study the various wavelengths of each color. You cannot measure darkness. A simple ray of light can break into a world of darkness and illuminate it. How can you know how dark a certain space is? You measure the amount of light present. Isn't this correct? Darkness is a term used by man to describe what happens when there is no light present."

Finally the young man asked the professor, "Sir, does evil exist?"

Now uncertain, the professor responded, "Of course as I have already said. We see it every day. It is in the daily example of man's inhumanity to man. It is in the multitude of crime and violence everywhere in the world. These manifestations are nothing else but evil."

To this the student replied, "Evil does not exist, sir, or at least it does not exist unto itself. Evil is simply the absence of God. It is just like darkness and cold, a word that man has created to describe the absence of God. God did not create evil. Evil is not like faith, or love that exists just as does light and heat. Evil is the result of what happens when man does not have God's love present in his heart. It's like the cold that comes when there is no heat or the darkness that comes when there is no light."

The professor sat down.

Rewriting My Stories

I moved to Austin, Texas, in 2017 because the loneliness and isolation in Aspen was overbearing for a newly single mother with a two-year-old child. I reconnected with friends in my entrepreneur tribe and girlfriends who had birthed their own babies since I'd last lived there in 2014.

During this time, I ran with my old crew of boho-preneurs and Austin hippies. We discussed manifestation, magic, miracles, money-making, and motherhood all day, every day. We went to ecstatic dance church on the weekends and sat in circles most nights conducting one or another form of ritual. That time in Texas was paramount to my acceptance that I was more powerful than I had ever imagined. But although the time in my Spiritual community felt great, it didn't elicit great results.

Although we talked about healing and magic, I saw that month after month I was filling my time with lots of positive self-talk but very little action toward my goals. I could cacao ceremony my way out of self-destructive behavior for a night or two, but I knew that I wasn't changing all that much. I looked transformed, but it was material, superficial, and rooted in pride.

Because I had just enough money saved to focus on being a full-time mom and part-time entrepreneur, I essentially became a professional student of healing. After spending most of the previous ten years in abusive relationships with sociopathic narcissists and even one criminal, I had to ensure that I would heal so that my son would live in a healthy home. That meant healing Mommy so that I could call in a proper stepfather to Hunter.

I'd already learned about manifestation through my mentor in 2010, who you'll read about in Chapter One. However, I'd become so exhausted with motherhood and my failed marriage that I didn't have energy left at the end of the day to manifest much of anything beyond sleep.

A natural-born leader, evangelist, and teacher, I dedicated my life once again to learning about manifestation; but this time, I didn't quit. For a period, I became a full-time teacher of the methods in this book either through private coaching, yoga-style retreats, and teacher training, and selling the Manifesting Secrets online ninety-day course. However, I eventually learned that my greatest contribution to world consciousness is through books.

I'm not the world's greatest manifester; I'm probably working on my bachelor's degree in manifestation, although my friends will kindly say otherwise. In other words, don't look to me for a life of perfection. I'm not currently married to the love of my life; in fact, I have no desire to be in a relationship right now because of the tremendous fulfillment I do have through my parenting and publishing company. I don't *yet* live in a $10 million dollar mansion, but I promise I'll publish a third edition of this book when I do.

But if you take nothing else away from this book, please know this: manifestation is a habit. It's a muscle that we exercise. With proper spiritual and mental nutrition, we can become manifestation masters.

This does not happen overnight, in 30 days, or even this year. For most of us, this will be a hard-won but deeply satisfying journey that starts off much harder than it will be once your manifestation muscles are well-toned.

Like me, you'll hit rock bottom a few more times in this life, God willing you have a bit of time left on this earth. And like me, you want to make the most of your time here. Perhaps like me, you believe that this is your only chance; there's no do-over nor reincarnation. For you this book is even more important; you want to make every second count in this life.

How Do I Get More Training?

When I built the Manifesting Secrets ninety-day program (www. manifestingsecrets.com), the goal was to help you make manifestation a habit that gets easier over time. This book is just one of many tools in that vault.

Manifesting Secrets is a loving compilation of audio, video, and workbook training from thirty experts in self-healing, neuroplasticity, epigenetics, energy work, relationship coaching, wealth, and other sciences and mediums that support manifestation, such as meditation, yoga, breathwork, NLP, and personal development.

The Manifesting Mind is your primer. In a few hours or days when you finish this book (because, of course, you won't be able to put it down), you're welcome to begin your 90-day Manifesting Secrets journey to begin developing the habits you'll need to make manifestation a lifelong practice.

At present, you're literally programmed like a computer to repeat past failures. It's subconscious; you don't choose these programs. Starting today, you will learn precisely how to rewrite your brain programs to avoid repeated failures in the future.

Email me at stephanie@manifestingsecrets.com with your questions, comments, concerns, and celebrations along this journey in *The Manifesting Mind* or to learn more about the ninety-day brain training.

I'm proud of you for choosing this powerful way to live.

Part 1

CHAPTER ONE:
Learning to Manifest

"Manifestation is the act of choosing your future through your mental intention. This intention is usually subconscious; your thoughts are like records on repeat with the programs you downloaded in childhood. Happily, with conscious rewiring we can interrupt programs we no longer want to make manifestation a blessing and not a curse."

Nearly ten years ago, I began actively studying the art of manifestation when my then-mentor shared with me his story of going from broke grocery store bagger with three kids to online information marketing millionaire. I was forever changed.

I took the most courageous action of my life that day. I said, "Look, I don't like where my life is. I understand that I somehow have chosen to be sick (with eating disorders), sad (in an unhappy marriage), sexually paralyzed, and miserable with my purpose in life: and today I am going to rewrite that story."

This man shared with me that I could change everything around me with the power of thought and self-belief. It sounded "woo-woo," cheesy, new-agey, unfounded, unscientific, and superstitious to me. But I said yes. I was dying inside and out: I really had no choice...

Armed with a mix of determination, discipline, tenaciousness, desperation, impetuousness, and a fair amount of faith, I quit my soul-crushing job in banking and began interning for two years under this mentor. I began my work as an administrative assistant in his company and became the Director of Marketing under his brilliant tutelage within 3 months. In the ten years since Jeff lit that fire in me, I've been a devoted if even sometimes foolhardy student of manifestation and the art of aligning mental intention to change my brain and, therefore, life.

In December 2017, a woman invited me to do a hot yoga class with her. I was living in Colorado with my son, then two, and we had a charming apartment in downtown Aspen that, although tiny, cost half my income; a savings account I drew from to live for as long as possible as a mom until my son was five and I went back to full-time work. I was making a little bit of money consulting businesses at that time, but I was barely making ends meet; motherhood is all consuming and, at the very least, two full time jobs. I was penniless faster than I'd liked to have been. Knowing I didn't have enough funds to cover even a simple emergency was unnerving. I felt irresponsible. In fact, it was irresponsible.

A Note On Poverty

Poverty is a recurring pattern you'll see in this book and it's one of the deadliest killers to a life of manifestation. Why? Wasn't Jesus poor? What about Mother Teresa? Isn't it harder for a rich man to enter the kingdom of Heaven than for a camel to pass through the eye of a needle?

First off, Mother Teresa generated millions from her books. Second, most of the great prophets and influential figures in history were quite wealthy. Why do you think that is? Why are we influenced by wealthy people? Namely, when you have money, you have a stage or reach. Additionally, people with the power of manifesting their realities through their minds generally learn to make more money. It is an express goal of mine that this book helps you to break the poverty patterns that limit your reach and power.

You see, when you're poor, struggling to make ends meet, and scrambling to put food on the table, your brain goes into a sympathetic nervous state of fight-or-flight. We'll discuss this at length later in Part One. However, it's important to establish a rule right now. **I don't believe that poverty is your**

destiny. I don't believe you can think clearly, strategically, or sovereignly when you're in struggle mode. It's neurologically near impossible.

Authors commonly use a model for writing a book that illustrates a Hero's Journey called "Save The Cat." It's a model originally used by film students when learning how to write screenplays and authors Jessica Brody and Blake Snyder have written brilliantly on the model. What you'll find is that most heroes or protagonists find themselves at a moment where "All Is Lost." You usually find that those heroes often have their big comeback moment shortly thereafter. I don't think it's wrong to have an All Is Lost moment. But my goal is that this book provides you with the understanding and power to claim your comeback moment. In fact, I daresay that you can hit rock bottom and still be relatively content with life. But your responsibility is to get out of rock bottom instead of making that your baseline.

Materializing One of My First Manifestations

Back to yoga class: I was slightly terrified. At that time, I hadn't done much yoga. I didn't know the difference between a Downward Dog and a Warrior Pose; and I certainly didn't know that so many versions of the posture! I'll never forget the first time a teacher said to assume the "warrior three" posture; I felt defeated; this didn't feel like a spiritual practice when I spent most of my time trying to catch a glimpse of what others were doing in the mirror rather than deepening my postures. I justified that yoga was boring and unnecessary for somebody like me who loved to lift weights, tumble, run long distances for fun, and who felt invigorated by otherwise difficult sports or exertion. However, I had seen many of my most powerful spiritual friends fall in love with yoga. In the back of my mind. I assumed that I'd use it restoratively if I was too injured to do anything "fun".

My friend insisted that I try hot yoga with her that day. And so, I gave it a whirl. It throttled me. I didn't realize that I would be dripping with sweat after the class, for starters. I drove home soaked, not having known to bring a change of clothes.

Although hot yoga throttled me, it also fascinated me. Previously I had attended vinyasa classes that compacted 70 postures or more into 60 minutes. It was overwhelming and I didn't feel surrendered to the practice; just stressed. In this hot yoga class in Carbondale, the pace was just slow enough that I could surrender into some of the postures that we held for thirty seconds or more. As I deepened my focus and attention, I could never have expected that I'd emerge from Savasana still tingling and trembling from activating energy that had been stuck for an indefinable amount of time. At once I knew why people talked about crying in yoga class. Activating and moving that energy in my body was an ecstatic, heavenly feeling. I wasn't encouraged to chant, worship Indian gods or goddesses, or partake in any rituals that conflicted with my then rather flimsy Christian faith.

Adding to the liberating and sensual experience in my body, the yoga teacher that day looked like a statue or a yogi master from a National Geographic magazine. He was tall, tan (I later learned that he's Armenian), lean, stunningly flexible, and comfortable in the practice. My friend coerced me into practicing next to her and, unbeknownst to me, I'd be practicing right next to the teacher, mesmerized by the postures and the elegant way he moved even though I was humiliated at my own practice in the mirror.

Several weeks later the yogi contacted me and we spoke on the phone for hours, like teenagers, exchanging book titles and discussing single parenthood. I was finally ready to allow a partner into my life who was kind, respectful, and loving. In Austin, I had been encouraged by my friend Sara Gustafson to write a list of all the qualities I yearned for in a partner. I kept it next to my bed and prayed for that partner, even visualizing what it would feel like to return to the mountains and feel less alone once I left Austin to return "home" to Aspen. By the time I met the yogi, I felt capable of loving and receiving love from somebody I'd put so much intention into meeting.

This man, Bel, was quite literally out of my dream. He was one of the first epic intentions I'd made that materialized. I set the intention that my partner would be another single parent. I visualized that he would be tall, dark and handsome! I intended that he would love the outdoors. I even manifested that he would rock climb; I had wanted to learn and had purchased a harness and shoes in anticipation of a partner one day teaching me this sport that called to me on a deep level. I manifested that my partner would love God and have a strong spiritual practice. I manifested a man who was a reader, a stargazer, and decidedly anti-television. Except for the fact that he didn't speak French, Bel was precisely everything I'd manifested on this exhaustive list I'd penned with intention. For two years I didn't go on a single date because I had standards and I refused to be wooed by anything less than my explicit intention.

Once weekly I visualized my perfect lover and imagined how we'd meet. I manifested that he would pursue me because I wanted him to harmonize my naturally assertive and masculine nature. I dreamed that my perfect lover would meditate and hold a spiritual practice with me. I actively put myself in situations where I might meet a man like this. I said affirmations in my mirror to nourish my own self-worth so that when a man came into my life, I would be strong enough not to lose myself again.

Most importantly, I manifested my perfect lover by becoming the perfect lover to myself.

But It Didn't End Well... Or Did It?

As abruptly as we fell in love, our home was broken. My lover's children hadn't yet processed their own parents' divorce and were struggling to accept a new family in their dad's life. When one of the children lashed out at my son, I called Child Protective Services and the family left instantly. My decision was based off:

1. Protecting my son
2. Protecting and preserving a healthy co-parenting relationship
3. Hope that my partner's children will get the psychological support they need.

I got saddled with a massive rental house that ate up what was left of my life's savings in no time. I went from as happy and fulfilled as I'd ever been to as lonely as I'd ever been, yet again. Worst of all, I didn't know at the time that this wasn't even close to my next rock bottom. I had no choice but to become a better manifester.

What Manifestation is Not

Well, this isn't a fun way to begin a book about manifestation, is it? Perhaps you were expecting nothing but yachts and private jets? Most likely, material results of this foundational groundwork will be in the third edition of this book. Although at the time of this second edition my life is considerably more exciting, wealthy, healthy, and happy than at the time of the first edition, I never want to lose this critical story of how I got started on my path to becoming, if I do say so myself, a positively epic manifester.

The rest of this book will detail the *hows* of manifesting. For over two hundred pages, you'll learn the mechanics of becoming a great manifester, all by yourself. But first, let's get these sob stories and some barriers to manifestation out of the way.

Barrier #1:
Living in The Past (Or Worrying About the Future)

To understand barrier number one to manifestation, let's explore the alternative to living in the past or worrying about the future: it's living in the present moment.

Why is there so much emphasis on the present moment in spiritual literature? The present moment or the "here and now" is quite miraculous. It's a place where we transcend time by being in a place where time doesn't exist. It's pure conscious awareness of the sensual and spiritual reality around you. It's not distracted by screens or to-do lists. It's a gift to yourself and those around you. In fact, scientists note the surest way to live in the present moment is while in intimate connection with others, where time disappears. There is no fear in the present moment; anxiety and worries disappear in the bliss of the now. It is between the past and the future where we are not haunted by the cares of either one. In the present moment, there is nothing but peace.

I have found that many teachers of manifestation or the law of attraction (which is just one of several components to manifestation) write nothing but happy, hopeful stories. Perhaps they live in the present moment more effectively than I! On one hand, this is brilliant. If you want to teach people how to live a life of hope, love, and healing, then don't focus on the bad. After all, what we think about expands. What we focus on, we attract.

However, I have noticed in the spiritual community, particularly since 2020's tragic events, that there is a tendency to put a smile on when things hurt. It's done with good intention; people want to vibrate at high levels of consciousness, including hope, joy, and non-attachment. However, that practice is also a cancer, causing more isolation and spiritual narcissism than it does help others to manifest their way out of hard times. To heal trauma and negative patterns or beliefs, we must have the courage to face them.

Barrier #2:
Toxic Positivity and Spiritual Bypassing

Putting a fake smile on top of pain isn't righteous, it's a form of toxic positivity or spiritual bypassing.

Denial of hard times isn't the way to get through them; it's a way of hiding, repressing, and manifesting that those hard times will persist. I didn't detail the sob stories from the beginning of this book to give them power; I detail them to let you know that it's okay to have pain. It's okay that you're suffering. You're not wrong to feel despair. As you see, I, too, have been in despair. Painful events won't go away; but the suffering can and will. *The Manifesting Mind* is your way to eradicate the suffering.

Few things really aggravate me, but I have a particularly impatient disposition toward people who lie about their lives and make it look "easy". Having spent many years coaching clients for anything from $2,500 to over $30,000, I belong to several private masterminds with other coaches in publishing, business, and marketing. We have a sort of running joke that if somebody spends too much time on social media trying to convince you of their happiness, that these are the following problems:

1. If they're successful and fulfilled, why the need to boast constantly about it?

2. If they're fulfilling some life's purpose, how do they allow themselves to invest that much time creating social media content?
3. Most photo shoots in private jets are fake; you can order them like you order a magazine subscription.
4. If you set the bar that high, you aren't touching your audience. You're shaming them. If you never have a bad day, you're a liar. I think this is why people like Jenna Kutcher have become such big influencers.

Call me old-fashioned, but I don't think it's supposed to be easy. It can be fun and even somewhat effortless; but easy is a shallow way to live. Deep evolution requires deep shadow work, deep humility, and deep dedication to the process.

There are few things that really grind my gears more than "spiritual bypassing" and toxic positivity. Spiritual bypassing is the act of denying pain to sound more spiritual. It's a downright selfish thing to do. It's entirely ego-driven and resembles spiritual narcissism. It makes the spiritual bypasser sound like they're "better" than everybody else, as though they're so "evolved" that they don't experience pain.

I don't know anybody successful who hasn't waded through dark waters while swatting at mosquitoes and pulling foot-long blood sucking leeches off their skin, figuratively speaking. I experienced seven years of fairly constant pain peppered with brief moments of hope and happiness to learn the lessons I've compiled in this book. It doesn't make me or you "bad manifesters" when we have shoddy days, weeks, or years. After all, my original publishing date of this book was 2021; inarguably following one of the crappiest years in human history for the mental health toll it's taken on the world, not to mention the vast poverty and suicides. Pretending everything is "okay" with spiritual bypassing completely goes against the laws of nature.

Barrier #3:
Paranoia (vs. Pronoia)

Because it takes between eight and twenty weeks to rewire a single thought, you'll hear me discuss affirmations frequently during this book. Although I ask my students in the Manifesting Secrets 90-day brain training to listen to at least two to five minutes of affirmations every day on various things such as wealth, health, love, confidence, and such, I typically have only one affirmation that I repeat every day, nearly the entire day.

If I get the joy of hiking alone in nature, I may repeat this affirmation for fifteen to thirty minutes repeatedly, carving the thought into my mind so that it will become automatic and part of my subconscious.

One such affirmation I used for a six-month period in 2020 was the phrase, "Everything always works out for me". It took a long time for me to believe this thought as global political upheaval made it hard to know which way was up. As a single mother, I was forced to take thirty-two out of a fifty-two-week year off of work to care for a five-year-old child who didn't have childcare nor school. Even once-close families suffered deep fracturing as riots, looting, vaccination scares and diverging opinions, and opinions on virus response ravaged us. During all this pain and trauma, a former partner filed multiple legal motions and Contempt of Court motions against me for, alas, my views on my city's response to the global crisis. In short, I posted in a local paper a skeptical comment about Aspen shutting down businesses and questioned what the effect would be on mental health. I posted on Facebook that we may be trading lives, not saving them. Based on these comments, the former partner initiated a hurricane of innocuous legal motions claiming that I wasn't abiding by the Colorado governor's recommended containment ordinances.

I'll never forget reading those legal filings after having spent six months feeling crippled and exhausted by taking care of a small child in the middle of a remote mountain town with no family around and my best single mom friends having recently moved to find environments with more opportunity and connection for themselves and their kids. It was one of the lowest moments of my life; I didn't think I had any energy for another crisis and suddenly my entire life would be consumed with legal battles over innocuous claims (primarily lies) from a vicious, nasty, well-funded and spiteful attacker. At that moment, I yelled at God. I felt abandoned. I doubted His existence in the core of my being. And if He did exist, in that moment I was seething angry with Him for standing back and watching me suffer beyond anything I'd previously conceived I could handle.

When I began my practice of believing that "everything always works out for me" with daily affirmations in those precise words, I initially took gentle, baby steps to believe that the legal drama would, soon enough, die down and result in my favor and protection. I then began to change my mindset about children being out of school, thinking about how I was getting my PhD in working fewer hours and making the same amount of money. It forced me to up-level my clients, keep coaching calls shorter and more efficient, and stop performing tasks that weren't on the critical path to my success.

As I repeated this phrase daily, "everything always works out for me", I also began to adopt the belief that every obstacle is my opportunity, that nothing happens to me, but everything happens for me.

This concept has psychological roots. It's the term Pronoia. Pronoia is the belief that everything around you is working out in your favor; the opposite of paranoia. Pronoia is powerful because it serves to help us attract what we want. Pronoia increases confidence, as you move through the world believing that the opportunities before you are part of a divine conspiracy to bless you. You more readily take phone calls, step on stage, showcase your talents, or pursue relationships you desire. Pronoia also increases manifestation; it causes our minds to concentrate on what good can happen as opposed to the neurotic negative thoughts of what bad things can happen.

When we focus on those good things, we attract more of them into our lives. In our brains we have a mechanism called the Reticular Activating System, or RAS. The power of the RAS is often taught in studies of the Law of Attraction, particularly when discussing the methods of visualization, mind movies, or vision boards.

The RAS is a bundle of nerves at the base of your skull that has several functions. Most importantly, the RAS serves to filter information between what you hear or see in your subconscious mind. Your RAS is your brain's gatekeeper; it decides what enters your subconscious or not.

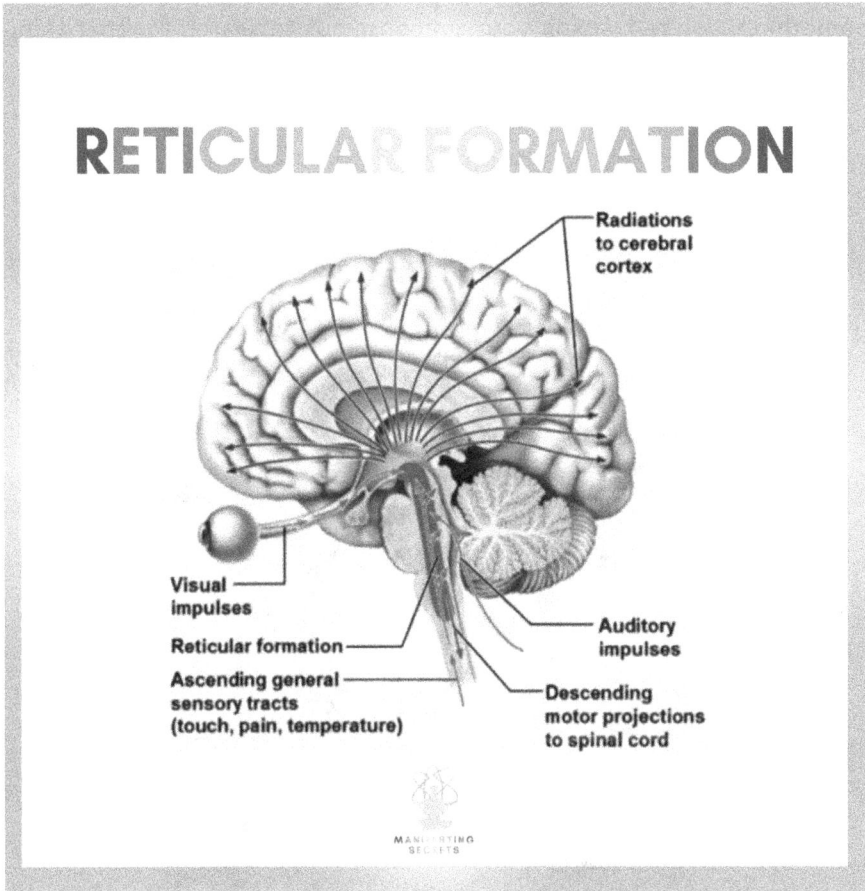

RETICULAR FORMATION

- Radiations to cerebral cortex
- Visual impulses
- Reticular formation
- Ascending general sensory tracts (touch, pain, temperature)
- Auditory impulses
- Descending motor projections to spinal cord

You have probably experienced your RAS in conjunction with something called the Baader Meinhof Phenomenon. Have you ever contemplated or purchased a new car and then suddenly seen this car all over the road? I used to want a Lexus LX350 and then I wanted a Tesla. Sure enough, I used to see the Lexus all over the roads, but then I saw the Tesla. There aren't any more Teslas on the road since I decided I wanted one. My subconscious mind got the alert that I wanted one, and my reticular was consequently activated to search for it in my environment.

Your subconscious mind is continually absorbing data from all over. In fact, your subconscious mind is said to pick up on at least eleven million bits of data per second (some say up to forty million), while the conscious mind picks up on roughly forty bits of data per second. If you want to become a better manifester, it's critical to set your RAS on alert for the data you want to consciously receive. When you focus your conscious mind on something, you're training your subconscious to search for that thing through your Reticular.

When you visualize what you want in manifestation practices such as vision boards, mind movies, affirmations or imagining what you want in a meditative state, you're showing your subconscious mind what to look for. Suddenly, you may begin to materialize the connections, relationships, experiences, and blessings you have been visualizing. These opportunities or manifestations were there the whole time; you've merely learned how to leverage your brain to materialize them. This is the power of *The*

Manifesting Mind. Manifestation doesn't have to be hard. It doesn't require lots of outside gadgets or superstitious practices. There's a science to the brain and body that add rocket fuel to manifestation; your greatest tool was factory installed by God Himself. These tools are gifts; they are reliable, provable, and powerful.

Barrier #4:
Attention to Negative Images

On the topic of visualization, there's also a shadow side to be wary of. What your brain perceives, it believes. It doesn't have the ability to distinguish between what you see and what is "materially real". For instance, if you observe a traumatic scene in a movie and feel scared (even though you know it's "fake"), your brain doesn't distinguish that you're not actually in danger or witnessing a horrific crime. It's possible to develop trauma based on what you see. This is why so many people turned off social media and televisions in 2020. Many believed that the constant images and reminders of sickness, disease, or financial collapse would cause them to manifest symptoms of these things. It soundly complies with these concepts of the Reticular Activating System. The Buddha once said, "what you imagine, you create" and popular Law of Attraction literature states, "what you focus on expands."

Can what we see on TV actually cause us to manifest poor health or physiological damage? Author of "Porned Out" Brian McDougal thinks so. If you watch or read a certain type of pornography, your brain identifies what you watch or read to be "real". If you consume pornography that differs measurably from what your lover or partner can provide for you, you may experience disappointment with your partner. You've calibrated your subconscious mind to expect a certain experience. McDougal states, "By messing up your brain chemistry, porn can cause erectile dysfunction, depression, delayed ejaculation, involuntary sexual fantasies, bad memory and concentration, emotional avoidance, poor relationship skills, sleep disorders, and a number of other problems."

Later on in this book, we'll dive deeper into specific visualization practices and how you can use them to lead a happier, healthier, and wealthier life.

CHAPTER TWO:
What is Consciousness?

"Love is misunderstood to be an emotion; actually,
it is a state of awareness, a way of being in the world,
a way of seeing oneself and others."
DAVID R. HAWKINS, MD, PHD

Just by picking up this book and opening yourself up to receiving the wisdom of manifestation, you have already begun to manifest. When we do something new, it interrupts the grooves and pathways that the synapses in our brains normally follow. It opens us up to neuroplasticity, rewiring the mind through mental intention. In a later chapter we'll talk about the two fundamental principles of neuroplasticity, but I'll attempt to repeat them a few times throughout the book so that they begin to become something of which you're consciously, effortlessly aware.

The first is this: **neurons that fire together, wire together**. What we do, think, or say repeatedly becomes a subconscious habit; it becomes part of the neurological structure of our brains. This means that we can manifest good things or bad things; it depends on the habits you create.

The second principle is: **what you don't lose, you lose**.

As much as this book is about learning new things, it's going to be about systematically choosing to lose the thoughts, behaviors, patterns, and belief systems that you do not want to carry with you into the rest of your life. Neuroplasticity is the ultimate form of freedom; nobody can take away the thoughts in your mind; and with those thoughts you control your entire reality and world. As you become consciously aware of your current patterns, programs, beliefs, and behaviors; you'll begin to consciously choose what to program or embed into your subconscious mind. Don't worry if this is a new concept for you; I'll go into greater depth in later chapters.

One of the first differences you will see around you as you begin to "wake up" and become more consciously in control of your life is by picking up on the vibrations and energies around you. This isn't about carrying a pendulum with you everywhere you go or "seeing spirits" and auras, but about awareness of energy.

Everything around us is energy. In fact, we have discovered through the science of quantum physics that all matter is made up of energy packets that are unbound by space and time. The energy that you emit ripples into the universe, as far as the stars, and returns to you.

People and objects around you are energy and emit energy as well. That's why you get "vibes" about things. Starting today, if you haven't already done so, I invite you to be passionately aware of how things "feel" around you. When you unlock this somatic or "body" awareness, you will begin to understand the world around you more. Your body, in fact, picks up on vibrations, energy, and the truth between the words much faster than your brain. If we trusted our bodies more, we would live in deeper harmony with ourselves, our choices with others, and with our manifestation power.

Deep into this journey of awareness, you'll begin to recall past stories, behaviors, beliefs, and words from others and how they felt. You will begin to see the truth around you; you'll understand the intentions of others with ease. In fact, you'll even become more sensitive to things such as music, clothes, art, and food. Everything, again, is energy. As you align with your own highest self and awareness, the world will look different. Sometimes, this takes the form of loneliness; you may begin to feel like you speak a different language than others or that you "don't fit in".

I trust that you may already have felt that way; perhaps you were the mouthy child who didn't fit in and hated to conform. Instead of lambasting or chastising these children, what if your parents or teachers had asked you why you felt this way? Children have less barriers to energy around them. They sense without abandon; they feel energy and thrive on high consciousness vibrations such as joy, love, and compassion. Children are some of our greatest teachers, if not the greatest.

Does The Brain Heal Itself?

The brain heals itself, but with your conscious direction. The brain is incredible, and you'll learn about concepts and tools in this book that you weren't taught in school, unfortunately. But although the brain is miraculous, it needs your direction. Without your conscious direction, your brain will stay stuck in old patterns that, after all, have kept you alive so far.

Your work in *The Manifesting Mind* journey will be to set the stage for parabolic healing to take place in your brain, which begets more health, wealth, love and happiness in your life.

Why doesn't this happen automatically if our brains are "so smart?" Why doesn't the brain heal itself without direction? Doesn't the brain "understand" that trauma must be healed? No. It needs your direction. You need to learn to speak the brain's language.

Over time, you've developed patterns, grooves, neurological networks, habits, behaviors, beliefs, and thoughts over the course of years or decades that have helped you to survive up until this point.

Your brain is interested in only one thing: survival, and doesn't want to change anything. Not if it's kept you alive so far.

That's why this journey is critical to your healing. Your brain doesn't heal itself. You're in charge of helping your brain heal by putting it into a safe place where neuroplasticity can happen.

For instance, your brain doesn't perceive that being codependent has made you psychologically miserable. It knows only that you're clothed, fed, sheltered, and living. So, it doesn't see the need to change anything.

Your brain doesn't perceive that you've been living just above the poverty line. It senses that you're getting by, and it doesn't seek your financial prosperity; it seeks survival.

Your brain thinks that the fight-or-flight state is a gift it gives you; your brain doesn't mold itself to give you a more peaceful existence. That is where you consciously direct your brain to heal.

Rewiring these patterns takes a bit of work, particularly in the beginning when you're identifying and creating intention around the new path you're choosing to consciously materialize. Although the journey to manifestation is front-loaded and can be arduous and tedious in the first few months or years, as you grow in consciousness, the trend becomes parabolic; it begins to snowball.

The interest of your daily manifestation habits compounds and becomes more effortless; like shares of a stock exploding, you will suddenly begin to reap the rewards of the compounding effects of manifestation.

If you're approaching manifestation because your life sounds like a country western song—your wife left you, your dog lost a leg, and you've been living on canned beans for a year... you don't want to hear that manifestation is front-loaded. However, I found that every coach with "false promises" about my own manifestation abilities left me feeling high and hopeful for a few days, only to return to my old patterns within weeks. Some of the most tortured souls I've met have invested too much hope in a one-stop shop retreat or Ayahuasca journey; when they return home from that journey, they have a false expectation that things will magically be easier. Although it's possible to have spontaneous healing, generally your deep-seated trauma requires dedicated work to heal.

In my experience, it's irresponsible for anybody who teaches manifestation to make these false promises without tethering their work to neuroplasticity and other scientific tools that make materializing a new reality easier, and even permanent. Tearing up your old maps, beliefs, habits, addictions, and stories can be exhausting, and at first it may be going too slowly. During these moments you must remember to be patient and anticipate that each day you're improving; each day you're happier; and each day you're more abundant. This concept will continually be emphasized and explained with increasing depth in future chapters.

To begin this process of rewiring your mind, it's important to have a foundational understanding of consciousness and energy. Focusing meditatively on your current mindset and consciously "rethinking" the thoughts you have will beget more peace, joy, love, and bliss in your life.

As a child, I grew up judging others endlessly and harshly. I name-called, lashed out, and berated people well into my thirties. I considered myself smarter than most people around me. I viewed others with judgment and a lack of compassion. In my personal commitment to elevating my own consciousness, I've found that instead of seeing people as flawed, ignorant, or jerks, I have found more compassion, seeing in others their own childlike innocence, understanding that everyone is doing his or her best, and seeing that the hurt people inflict on others is a reflection of the pain they themselves possess. This helps me lean into my relationships with more intimacy, grace, silence, eager listening, reflecting, and ultimately helps me manifest more abundance from the world around me.

Our thoughts manifest our emotions, our words, and even the material world. When our thoughts vibrate higher on the scale of energy consciousness, we'll be more at peace with ourselves and the world around us. We exponentially increase our joy, happiness, love, and even enlightenment when we fill our minds and environment with things that promote higher energy consciousness. For me, this means that five years ago I nearly altogether quit listening to mass media "news", radio, music with negative lyrics, or even glancing at tabloid headlines while in line at the grocery store. I understand that so many of these information sources are designed to lower my vibration and put me in a state of fear or anger.

The Map of Consciousness™

Now let's look at how this works from the scientific level. One of my favorite teachers, Dr. David R. Hawkins, has written extensively about how the energy of positivity and negativity affects the world around us. In his book "Power vs. Force", Hawkins illustrates that humans live at remarkably varying levels of "consciousness." He has calibrated these levels on a logarithmic scale called the Map of Consciousness™ that rates anything from emotions to people to book passages to songs between levels 1 to 1000. In the search for enlightenment, the goal is to reach as close to 1,000 as possible.

MAP OF CONSCIOUSNESS
• INSTITUTE OF ADVANCED SPIRITUAL RESEARCH

God-view	Life-view	Level		Log	Emotion	Process
Self	Is	Enlightenment	⇧	700-1000	Ineffable	Pure Consciousness
All-Being	Perfect	Peace	⇧	600	Bliss	Illumination
One	Complete	Joy	⇧	540	Serenity	Transfiguration
Loving	Benign	Love	⇧	500	Reverence	Revelation
Wise	Meaningful	Reason	⇧	400	Understanding	Abstraction
Merciful	Harmonious	Acceptance	⇧	350	Forgiveness	Transcendence
Inspiring	Hopeful	Willingness	⇧	310	Optimism	Intention
Enabling	Satisfactory	Neutrality	⇧	250	Trust	Release
Permitting	Feasible	Courage	⇧⇩	200	Affirmation	Empowerment
Indifferent	Demanding	Pride	⇩	175	Scorn	Inflation
Vengeful	Antagonistic	Anger	⇩	150	Hate	Aggression
Denying	Disappointing	Desire	⇩	125	Craving	Enslavement
Punitive	Frightening	Fear	⇩	100	Anxiety	Withdrawal
Disdainful	Tragic	Grief	⇩	75	Regret	Despondency
Condemning	Hopeless	Apathy	⇩	50	Despair	Abdication
Vindictive	Evil	Guilt	⇩	30	Blame	Destruction
Despising	Miserable	Shame	⇩	20	Humiliation	Elimination

MANIFESTING SECRETS

Below 200 is considered a "negative" force. Sadly, Hawkins posits that approximately 85% of humanity operates at a level of consciousness below 200, which is "negative" force.

Interestingly, negative patterns are associated with sickness and positive patterns are associated with health. Everything from books, food, water, clothes, people, animals, buildings, cars, movies, sports and even genres or styles of music can be calibrated at different levels. And sadly, a fair share of today's music calibrates at levels below 200; meaning it propels lower energy of consciousness; and behaviors of the like.

But here's something shocking and exciting: **you have the power to counterbalance the weakness of others** in an exponential way. Your energetic vibrations from higher places of consciousness counterbalance the negative vibrations of others. Positive vibrations are substantially stronger than negative, weaker vibrations. Here's how Dr. Hawkins breaks that down:

* *1 individual at level 300 counterbalances 90,000 individuals below level 200*

* *1 individual at level 400 counterbalances 400,000 individuals below level 200*

* *1 individual at level 500 counterbalances 750,000 individuals below level 200*

* *1 individual at level 600 counterbalances 10 million individuals below level 200*

* *1 individual at level 700 counterbalances 70 million individuals below level 200*

* *12 individuals at level 700 equal one Avatar (e.g., Buddha, Jesus, Krishna) at level 1,000*

How can we calibrate at higher levels of energy to positively affect the world around us? How can we as individuals shift the energy of our homes and towns for more positivity and less anger, stress, or apathy? You guessed it: it's through our habits.

Each time we meditate we elevate our vibrations. Each time we choose silence and awareness before reacting to things that trigger us, we elevate our vibration. Each moment we breathe deeply and envision aggressors as hurting children lashing out, we form the habits of higher consciousness. After all, one of the highest vibrations to embody is that of non-judgment and nonattachment.

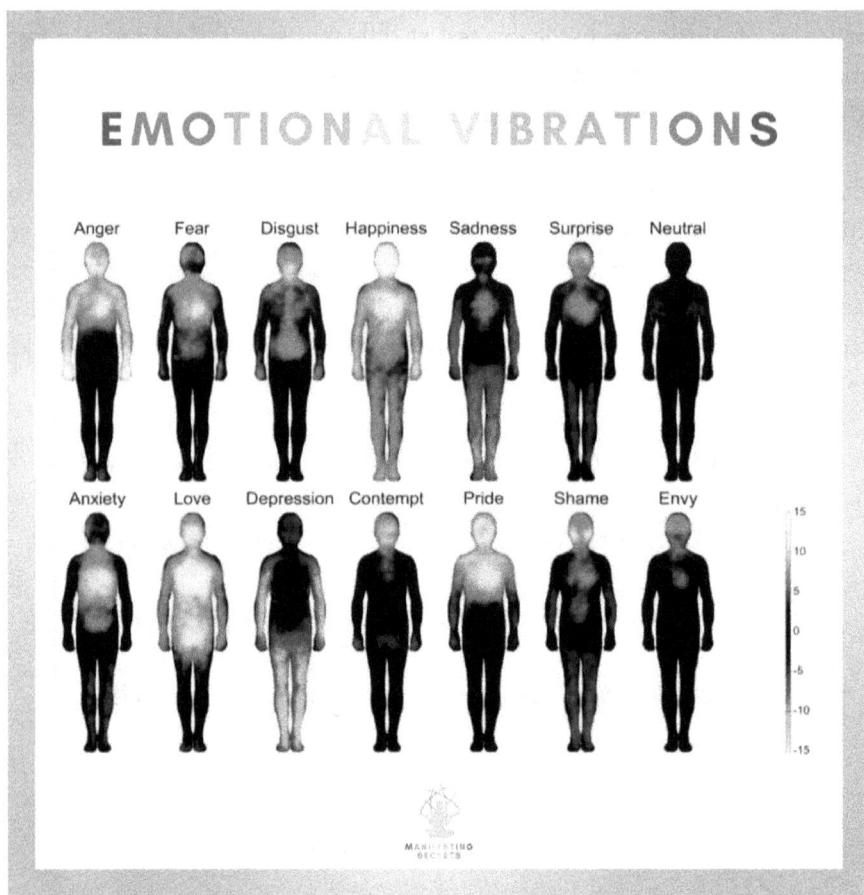

Society calibrates at a mean level of approximately 204, or just above negativity. Earth's great minds calibrate around 499-510. Unconditional, unchanging, permanent love is categorized at a 500. This isn't "love" for beer, football, or even your spouse. In fact, only 4% of the population ever attains this level.

At level 540 is Joy. At this level, love is truly unconditional and is experienced as inner joy that does not shift with the turn of events. Only .4% of the global population is at this level, which is also asserted as the level at which the third eye opens; physically known as the Pineal Gland.

At level 600 the popular book "A Course in Miracles" comes in. Slightly above "A Course in Miracles" is the New Testament of the Bible, which calibrates at a 640. Provided the book of Revelations were removed, the

New Testament would calibrate at a 790. The Old Testament, however, calibrates at only 190.

According to Hawkins, for the Bible to be "above negativity", it would have to exclude the entire Old Testament except for Genesis, Psalms, Proverbs, as well as Revelation.

Interestingly, the Lamsa version of the Bible which is a version of the Bible translated from Aramaic (Jesus' language), sans the Old Testament (except for Genesis, Psalms, and Proverbs and Revelation) would calibrate at a level 880.

At level 790 is Manifestation, which is defined by Hawkins as "the emergence of potentiality out of essence into manifestation." In the 6th century AD, Buddhism calibrated at a level 900 and has deteriorated much less than other religions since that time. The Bhagavad-Gita is calibrated at level 910.

The esoteric teachings of Jesus Christ which survived through Gnostic texts calibrate at a 930. The Upanishads calibrate at a 980 and at a level 990 is the statement, "There is no cause of anything."

Now please study the Map of Consciousness chart above to get a quick start guide to energy boundaries in your life.

Dr. Joe Dispenza says in his book, *Becoming Supernatural* that emotions are "energy in motion". Assess the emotions and feelings you've had today. Take a moment to evaluate those feelings or emotions next to the Map of Consciousness™ chart and as you read the rest of this book, begin to strategize how you will elevate your own vibration to bless yourself and the world around you.

CHAPTER THREE:
The Law of Nature and the Yin-Yang

"Yin and yang, male and female, strong and weak, rigid and tender, heaven and earth, light and darkness, thunder and lightning, cold and warmth, good and evil...the interplay of opposite principles constitutes the universe."
CONFUCIUS

Now we'll focus our minds' eyes on the hope and promise that comes from what is referred to as the First Law of nature. This is the law of yin and yang. Simply put, darkness follows light, but light follows darkness. You can be Christian, Buddhist, agnostic, or otherwise, and still accept this concept and the sound wisdom it brings.

My grandmother once suggested that I could be in danger of hellfire for doodling yin-yang's in a notebook and "worshiping idols". She's uncomfortable with the Buddhas on my wall, too, which I've recently chosen to donate to a local yoga studio.

I don't think my grandmother was a crazy fundamentalist. She pointed out to me at a young age that keeping my eyes on Heavenly things were critical to my spiritual life. What she missed was the marketing! You see, I didn't see any problem with listening to popular music, watching popular movies, or even dabbling in various religions until a couple years ago.

Grandma wasn't right, in my experience, to request that I remain sheltered. But she was spot-on to point out that diverting our attention from the heavenly war around us is a sure path to destruction. I'm living proof. In recent years I've become more aware that this world is far more evil than I could have ever imagined as a child. I see now that my grandmother didn't want me to be a delicate flower, but a spiritual warrior.

On that note, consider the things that distract you from your spiritual path. Music could be one of them. Incessant TV or Netflix watching could be another. Football is an obvious opiate of the masses. Now consider the broken homes, the child trafficking, and other tragedies around you. I am of the conviction that while some "self-care" is necessary, so, too, is being so fueled by purpose that you don't choose to spend ample time in this dopamine-abusing hedonistic space. The teachers many of us look up to because of their wisdom lived lives of great discipline with harmony between their humanity and spiritual selves.

Another aside before diving into this concept of harmony and the yin and yang is a note on fundamentalism, since I mention my grandma's chiding me regarding the yin yang symbol. For years, my historical, philosophical, and spiritual studies were stunted by a sort of indoctrination from my fundamentalist Pentecostal upbringing. Many people in religious communities were brought up to think that academic study, skepticism, and truth-seeking could lead to damnation; I was one of them. As a devoted learner and skeptic above all else, this closed door paralyzed me. I felt the suffocation of only being allowed to read one book for the rest of my life. At some point, it caused me to reject all spiritual and personal development altogether. It gave me the feeling that I could never do it right. I've had other family members lecture me for hours on the dangers of yoga. I bring these incidents up because they are haunting, frightening, and fear based. If your faith is eroded and rebuilt 1,000 times in your life because of your honest pursuit of truth, I suggest that God might be supremely pleased at your authenticity and humility to evolve.

There's something downright breathtaking about my grandmother's immovable, unshakable, unwavering, confident faith in her truth and the way she chooses to express it. The older I get, the more boundaries I, too, put up to guard myself from what I consider to be evil in this world. I enjoy the fruits of a life of discipline and faith. My grandmother is old-fashioned and definitely not a scholar, but you'll never meet a woman who has more vivid dreams and more accurate intuition. Her prayers can be felt from 1,000 miles away, I can attest. And, too, there's something equally breathtaking about her hot pursuit of truth in her way. My hot pursuit of truth is another way; taking in wisdom from all cultures and trying to understand the foundation of all faiths; which has, in turn, brought me to an unmovable faith in Christ.

I honor your journey. I don't judge you for embracing or rejecting any one creed or faith. My walk is unique to me alone as are my thoughts. When you are tempted to judge somebody else, consider this phrase I repeat often to my son: "We are all fundamentally good people who make periodic bad choices. We are all doing our best." This helps us vibrate on a level of love, acceptance, non-judgment, and other levels of consciousness that are regenerative and impactful to all of humanity as a blessing of our energy.

Your current life is the sum of every thought you've ever consciously had, or those which have been implanted through subconscious programming. I appreciate how many positive spiritual messages and songs my grandmother and other spiritual teachers encoded in my heart and mind over the years. Your body, relationships, health, and even your income are reflections of the blessings or curses that you received as a child.

Now we're going to step into one of the most powerful forces of creation within you: your harmonized divine feminine or divine masculine nature. Whether you're a man or woman, you have both masculine and feminine energies. The feminine energy, or yin energy, is the force that literally creates life. Our nurturing sense, our intuition, even compassion all stem from feminine energy. We like to define masculine and feminine energy by the terms "yin" and "yang".

The Yin-Yang Symbol

The yin-yang is an image that's at least 2,720 years old. It's a Taoist symbol known as Tai Chi. Tai Chi is the symbol for life. In Chinese characters, the yin-yang is the image of a house on a hill. On the left side is the yin; the yin is represented with the night and the moon. On the right side is yang, the sun and daytime. Yin energy is passive and cool; yang energy is active and hot.

Wu-Wei is the balance between yin and yang. Wu-Wei is also known as "the way of life" or "purpose." Our purpose is to balance the yin and yang. One great way to identify yin and yang is to imagine a conversation. If I'm talking, it's yang energy. If you're talking, it's yin energy. If we're both talking, it's chaos. Hence the importance of balancing the two sides.

Yin-yang applies to emotional, spiritual, relational, and professional needs. Your job is to find Wu-Wei, or the balance between yin and yang. Pierre Chardin once posed the question:

Are we human beings having a spiritual experience? Or spiritual beings having a human experience?"

Without yang energy, we likely won't earn enough of a living to have freedom. Without yin, we lack rest and connection to God and Nature. Yin is said to be the recipe to going quantum. You don't need yin development; it's part of your spiritual nature. To answer Cardin's question, we are spiritual beings having a human experience. But while tethered to these bodies, it's necessary to find food and shelter; to be in our "yang" energies on this earth or plane of existence.

Yin-Yang and Yoga

Yoga is one of the most powerful ways of engaging manifestation in your life. While setting an intention, speaking your intention, and moving with that intention, you employ all three executive functions of the brain; making it easier to rewire old thought patterns you seek to improve.

While this isn't a book on yoga, we'll quickly identify a few core principles that may help you on your manifestation journey.

Yoga is not about postures. It's the yoking of an individual with the divine through breathwork, meditation, and movement. If you ever hear me speak of yoga, I'm not talking about yin, vinyasa, or postures, called "asana." Rather, when I use the word yoga, I'll be referencing this threefold practice of meditation, breathwork, and movement.

The traditional teachings of yoga identify 72,000 energy meridians called "Nadis." The main tube is the Shoshuna/cord in your spinal column. This is why yogis work on lubricating their spines, stretching and twisting, and otherwise "exercising" the spine so that they can sit still with erect posture in meditation. Asana or postures aren't a workout; they're movements designed to ultimately bring you to stillness for meditation, where your brain is at its highest potential for neuroplasticity or brain-rewiring. I'll define neuroplasticity at length in Chapter 11.

It is believed that you have a clearer channel to the Divine and manifestation when you meditate with an erect spine. There are numerous benefits to the body and brain while meditating or praying in this position.

You often hear of "activating" Kundalini regarding manifestation. This can feel like electricity in your pelvis or lower back or pelvis regions. Moving my body in yoga asana while dedicating my breath to focusing on each posture as an act of celebration to my strength and discipline is a way I activate the kinesthetic executive function of my brain, which is required for turning intention into materialized manifestations. You may hear Christians or other religious groups say that yoga is a sinful practice. However, many of the postures I learned came from a Christian teacher who is passionate about brain science and manifesting our intentions through thought. She teaches yoga postures as a form of activating your body's divine healing program. That's right, you have a healing program; it may just be dormant. After reading this book, you'll understand how to reawaken the magnificent ways you have been designed by God to heal. You will also hear me discuss meditation at length in part 3 of this book when I give you a seven-step process for igniting manifestation through meditation, which you can sub for "prayer" if you so choose.

Yoga For Emotional Regulation

Have you ever felt out of balance in your emotions? Usually, this is because you have out of balance "yin" energy; it's common in most relationships to have this temporary roadblock. Yin energy is passive, introverted, yielding, soft, intuitive, and nurturing. Yang energy is active, extroverted, dominating, hard, logical, and initiating. A great way to assess whether you're "out of whack" is to first look for excesses.

Here are what the excessive traits of yin and yang energies look like:

* Needy
* Confused

* Martyr
* Depression

Here is what the excessive traits of yang energy look like:

* Forcing
* Rigid

* Domination
* Anxiety

In your Manifesting Secrets brain training, you'll find many worksheets, reading and videos, designed to help you identify your own balance of yin and yang energies with prompts and exercises to help you bring more harmony to your body, relationships, wealth, health, and happiness.

To enjoy a guided meditation for free today, please visit www.manifestingsecrets.com/bonuses where I have prepared a meditation for you to listen to today.

Yin and Yang in Spirituality

> *"You have heard that it was said, 'You shall love your neighbor and hate your enemy.' But I say to you, love your enemies, bless those who curse you, do good to those who hate you, and pray for those who spitefully use you and persecute you, that you may be sons of your Father in heaven; for He makes His sun rise on the evil and on the good, and sends rain on the just and the unjust.*
> MATTHEW 5:43-48

Yin and Yang can also be applied to your spiritual or overall non-body state. When you're in darkness, you fight shadows, scarcity, demons, and past traumas or generational curses that generally make the darkness even darker. When you're walking in the light, you vibrate high and attract more of what you want into your life.

My grandmother reminds me that even "good people" have bad times. She reminds me that the Bible says, "the rain falls on the just and the unjust." There will always be dark moments to harmonize the light. They teach you and temper you. Dark times prepare us for even greater times of abundance.

You aren't "wrong" for having bad days. Most excitingly, it's the bad days and contraction that make way for even greater expansion. Like a wineskin, when you expand in your personal development, you may shrink and contract as your body catches up to your spiritual expansion, but you will never go back to the "regressed" form. You will keep expanding.

Look, if you have never had a bad day, get out of my book right now. You won't like it here. We're going to be raw, honest and vulnerable. This is a book of spiritual and scientific tools for real people. If you're spiritually bypassing bad days, you won't glean anything from the content. Frankly, I'm not sure I want you here. If you're willing to be honest, transparent, vulnerable and even pissed off with me, we can do this.

CHAPTER FOUR:

Turning Sludge into Sugar

"The goal of a manifester isn't perfection;
it's consistent manifestation habits."

I once heard Ryan Holiday describe turning "sh*t" into sugar through a stoic life in his book "The Obstacle is The Way." It stuck in my head during those weeks after my lover left my home and I longed for him with a sort of paralyzing desperation. It was like losing a limb. If you've ever been deeply in love, you understand what I mean. I desired the bad to be switched to the good; hence why the phrase "turning sh*t into sugar" stuck in my head. I was determined to turn my life around, as I have done before, and I was going to do it again. This time I would know more about my powers of manifesting so I could focus on the good. I had to hit rock bottom, though, before I could manifest my way up.

Be encouraged with this: the better you get at manifestation, the quicker you recover and rise from darkness.

In 2018, I fell into a place of darkness, despair, and victim mentality. Upon seeing that my partner's kids weren't ready to share space with me and Hunter, I removed my partner and his family from the home that had once been partner's so that I didn't have to move my son twice in six months. An enormous rent was now my responsibility to pay. My coparenting relationship began to fray, which terrified me. I prayed for two years for an epic love to come into my life, and it felt like I would never meet another man like my partner again. I was scared of being alone, of running out of money, and of constant threats from another litigious, vicious former partner from my past. My first postpartum business failed miserably, and I lost a sizable amount of money I had invested into building the business and lifestyle around it. I was penniless; going into credit card debt to make ends meet. I didn't know what I would do with my life and began to question my life's purpose. I had thought that I was a teacher, a leader, and an inspirer. Now I just felt like a sham. Everything I'd manifested: the lover, the business, the family, and the safety and security; everything collapsed. And then, so did I.

One dim December morning, I kicked a patch of snow to dislodge a toy my son had left there the previous week. I didn't realize that the patch of snow had solid ice for several inches beneath it. I instantly bruised my joint beneath my right toe and cracked a bone in my right foot. To make matters worse, as I babied the foot over the next several days, I somehow slipped a disc in my back. I was in crippling pain and could barely walk. When your stomach, eyes, head, chest, and other parts of your body "react" to things around you, that is the body giving you signals about the world. If you are unwilling or unable to fully process and release emotions, they will get "stuck" in your body, particularly in the chakra system. I held a lot of tension in my solar plexus and most likely slipped the disk in my lower back due to the trauma and weight I carried in my stomach and heart, the third and fourth chakras, respectively.

I learned that different parts of the body are associated with different experiences, such as trauma. For instance, an expert in Five Elements acupuncture and Chinese medicine found that because I was scared for my son and my ability to care for him, the right toe was injured because that's the maternal line, although on the "masculine" side of my body. It sounded crazy to me, but also fascinating, and spot on.

My business at the time hadn't brought in revenue since Bel moved out; I was suffocating on self-pity and my own sadness, unable to work with a clear head. I was plagued with busywork; a veritable hamster on a wheel; hobbling, no less.

Sometimes I thought the pain to my ego from my failing business hurt the worst of all. I got angry. I acted like an absolute witch to everybody I met. I began to destroy my reputation among my small Colorado tribe. I lashed at Bel and berated him because I hated myself and my life. I hated him for being gone. I hated the prospect of "looking" for love again. I was looking for a victim to blame everywhere I went. I'd blame customer support agents, grocery store baggers, moms on the playground, nobody was safe around me. I pointed the finger at Bel and began to chastise him for something that we had both created, chose, and materialized.

Losing My Personal Power

In playing victim like this, I lost my personal power. In fact, it was a shadow I'd always struggled to own and was now allowing myself to run free for hopefully its last hurrah. I didn't have the energy to stop it. My inner child was seething mad and throwing a temper tantrum. I verbally assaulted everyone around her because I was projecting my own pain and sadness. Hopefully letting it out and recognizing what I needed to finally heal myself. Victim mentality is something I've struggled with since I was four years old. I was a true complainer; the quintessential attention-starved oldest child of four apparently much more needy children. To get attention from my time-strapped mom, I complained to anybody who'd listen.

The sad part was, I wasn't even attention starved. My family doted on me for miles around. My grandmothers and great grandmothers adored me. My grandfathers, uncles, cousins, aunties, and great relatives from all over the country showered me with gifts; sent me cards on every holiday and showed me I was nothing if not the new incarnation of the messiah himself. Yet, somehow, I had it "in my head" that I was a victim.

What happens when you play victim?

* Your body shrugs, slumps over, and manifests the feeling of being "beat up."

* Your life becomes less successful; instead of confidently making strides in your career, you'll always look at somebody ahead of you on the pay scale or ladder and claim that they didn't earn their success or even that they trampled on yours.

* You lose confidence; you begin to tell yourself that nothing works out for you.

* You become scared or fearful; people are out to hurt you.

* You sabotage relationships by finger-wagging, blaming, judging, and generally not listening to or holding space for others.

When you play the victim, you sabotage your success.

One of my best girlfriends came to live with me for a few weeks that holiday season. She was mortified by my psychological and emotional state. She had no pity for the failing business; she told me to turn my back on it or, as Kevin O'Leary so poetically says in The Shark Tank, to "take it out back and shoot it." She pointed out that my relationship with Bel had gone from sovereignty to codependency yet again; the very thing that plagued my marriages (yep, two failed marriages already, at that point). She told me that my physical injuries would persist and even worsen if I didn't get out of the psychological and emotional pity party I was in.

How right she was. I had manifested the pain. I had manifested the broken foot. I had manifested the breakup with my lover. I had manifested the family tragedy, the codependency, the poverty... and so much more. Although there were many factors at play in facilitating my ostensible demise, I had many opportunities to choose paths that perpetuated the pain. Although I had indeed been "victimized" by vicious, litigious individuals, I had also reacted to these attacks in a way that dug the hole deeper and deeper.

Upon realizing and accepting this truth, I was ready to heal that pain and go to the deepest root. My broken foot was simply a symptom of a larger problem. My poverty was a symptom of an even bigger problem. My failed relationship, yes, this was a symptom. The loss of my best friend was a symptom. None of these things made me miserable. I made myself miserable with one sickness that elicited all the unhappiness in my life: my self-sabotaging thoughts. In truth, I could look back and call that the happiest time of my life. Why was I so sad? I chose sadness. My thoughts created suffering when I could have felt the pain and adjusted my life and mindset to return to happiness and gratitude again. Let's learn more about how I healed, and about how you can heal, too.

CHAPTER FIVE:
The Brain Experts Who Saved My Life

"Science is the contemporary language of mysticism."
Dr. Joe Dispenza

At that point in 2018, I was probably a bit maniacal. I wasn't just having a tough month or season, it felt like I was heading into yet another miserable year in 2019. I had whiplash from being so happy with my lover and suddenly suffering so deeply.

Thankfully, my friend Susan worried about me and asked me to visit her mentor at the Neurofunctional Institute (NFI); a radiant, brilliant neurofunctional healer named Lauryn Gepfert.

Susan had mentored under Gepfert for a year and was already taking clients as a healer in her own right; the duo and their team at NFI were healing quadriplegics in swimming pools with their thoughts; they were healing people with neuroplasticity or rewiring the mind through mental intention. That's right, they were making the lame walk. Jesus-style. As a former youth minister, I was beyond intrigued. I was obsessed. In fact, I truly believed that I, too, could experience healing. I hadn't experienced it yet because, well, perhaps I didn't have a strong enough faith. Perhaps I didn't have faith because I didn't know *how* to heal.

I chatted with Susan often about her work over coffee or kombuchas, but I hadn't actually experienced it hands on. When Susan magnanimously set up an appointment for me to chat with Gepfert about marketing their incredible business, I could barely get up the stairs and sitting in a chair was impossible; I was in so much pain that I had to take breaks from driving my car to get to the office only seven miles from my home for the pain it brought my foot and spine to push the pedals.

Gepfert and I didn't discuss marketing that day. We weren't meant to. She and Susan subconsciously chose to bring me into that office to heal me. Within moments of Gepfert entering the office, she had me lying face down on a massage table and she was repeating words and phrases that were designed to rewire the pain programs in my mind and activate self-healing. I don't know how long Lauryn worked on me that day. I do recall leaving her office feeling less pain because she believed I could heal right there on that table. However, I didn't heal right there on that table.

One of the most critical teachings in Gepfert's work relies on setting intention in the conscious mind to heal the body. This is the same process we'll use to manifest the lives we want. When Lauryn caught me explaining my past traumas or blaming my bum foot or back on incidents, she mindfully urged me to stop living in the past. "We don't look back, we sensory stack," she teaches.

Lauryn and Susan taught me that although I believed in my ability to heal, I didn't believe in myself. I wasn't choosing to heal, and I didn't give myself permission to heal. After my work with Lauryn on that massage table, the healers urged me to stop putting my healing in Lauryn's hands, in the hands of my bank account, or in the form of a ring on my finger or a baby in my womb. Lauryn showed me that with mental intention, healing language, and healing movements; I was my own healer.

When I got home, I hobbled into my living room, cursing myself for not having recorded Lauryn's words and incantations. I laid on my back and repeated a few that I could remember. "Drop. Drop. Drop," I commanded my back and foot. I visualized the bones welding themselves back together and the body bringing healing from every corner, crevice, limb, and appendage to the injuries. My son would be coming home in a few days. I didn't know if I could make another week alone with him; I could barely lift him, and it didn't feel safe to have a child in my own crippled care. "Drop. Drop. Drop..." I continued.

"Back, you don't have to be in pain. You can relax now. You are safe. Foot, you don't need to throb. This body has everything you need to heal right this instant. As I prayed, meditated, and spoke words of healing over my body in the second person, as Lauryn had directed me to do, I periodically wiggled my toe. I envisioned a light coming in from the heavens and filling up my whole body, activating my natural self-healing capabilities. I relaxed for four hours, just sitting and meditating, crying a bit, releasing my sadness, and putting the suffering into God's hands. "Body," I said, "you don't have to hurt. I recognize your pain. It's okay to take all the time you need to heal. I will nourish and protect you. You are free to heal. You have permission to heal."

I repeated the words Lauryn used until I believed them, until my critical, left brain turned off and I began to live in the magical quantum yin energy that I detailed in Chapter Two. My mind turned off and I was in pure healing bliss. That day, I got excited for healing. I felt confident in my ability to heal. That night, I was reborn into a powerful manifestor.

After four hours, I arose from the acupuncture pillow I had been lying on. I held my breath as I lifted my lower back to pull out the pillow from under me. My chest caught in my throat as I curled my back to stand erect. My heart stopped when I stood up with ease. I had manifested something... I was pain free. I walked around my home. Back and forth. Back and forth. I paced, I cried, I worshiped God and shouted yes, yes, yes over and over. I believed I was healed.

Just to prove it and surprise Bel, I shot a video for him of myself doing a backbend and back walkover in my living room. He could hardly believe his eyes; that morning I had struggled and cried merely trying to get out of bed. I had experienced a true miracle.

In just five days after the healing session with Lauryn and Susan, I was back in yoga, balancing on my formerly broken foot. My mind was truly blown, and my heart melted with appreciation and self-love. Lauryn and Susan reminded me of what the Bible taught; this verse had always come to mind when I thought of what a miracle is.

"Very truly I tell you, whoever believes in me will do the works I have been doing, and they will do even greater things than these, because I am going to the Father."
JOHN 14:12

Lauryn and Susan were creating miracles. More critically, they were teaching others how to be miracle-workers with the simple act of mental intention. In the coming months, Susan spent many generous hours teaching me about how I could heal my brain and showing me her method of healing that incorporated a multitude of modalities from Ayurveda to personal training. Susan allowed me to interview one of her clients who had healed a tumor in his back with mental intention and self-healing language.

While watching Lauryn and Susan heal others starting with mental intention, I began to believe in myself in a deeper way than I ever had before. Lauryn didn't just help heal quadriplegics. I learned that she also taught powerful CEOs and entrepreneurs how to amass more wealth through mental intention. She helped the elderly, who she calls the "ageless actives" versus the "actively aging," learn how to reverse age with her work. She helped heal individuals with eating disorders. In fact, the list is profound.

Anything you want to heal can be done with your mind. And for all the superstition, Tarot cards, and rituals I'd tried to manifest; this put everything else to shame.

You are your own self-healer. Manifestation is about healing the mind from negative programs. The power is within you. The power is you.

Now let's dive deeper into some more of the manifestation miracles I learned over the next fifteen months.

CHAPTER SIX:
The Science of Manifestation

"Neuroplasticity is simply the act of rewiring your mind through mental intention."

Fundamentally, manifestation is:

Always happening whether we like it or not.

So... we're better off consciously choosing what to manifest.

Let's learn how to incorporate actions that align your brain through movement of your body, breath, spirit, and thoughts; all within real scientific principles proven to help us become happier, healthier, wealthier, and more in love with life... and ourselves.

Your Life is a Result of Subconscious Programming

Since you are always manifesting, let's choose together to direct our thoughts, words, and actions toward a goal we consciously choose instead of what you've been "programmed" to think, do, say, and receive.

You see, it has been discovered that nearly everything in your life, before you learn to harness your manifestation power, is a result of subconscious programming you had little to no control over. We'll learn about that a little later, but suffice it to say, the reason we must learn to consciously manifest our lives is because if we aren't consciously choosing our lives, we are entirely controlled by our subconscious programming. This fact may sound dire, salacious, or even exaggerated. However, scientists have found that ninety five percent of our choices come from neurological programming we downloaded before the age of seven. We are the result of the neurological programs, religion, education, trauma, abuse, fear, disease, laws, legalism, and domestication of our families and environment. Unless we consciously choose to rewrite what was hardwired into our brains when we were children, we will continue to repeat the same patterns; for better or worse.

Your pattern may dictate your propensity to worry, fear, or feel anxiety. You may say, "I'm a worrier" or "I'm bad in relationships." The truth is that you are failing in the areas of love and peace due to previous neurological programs you've downloaded; you're the result of maps in your brain. Remapping those negative thoughts and, consequently, actions, is what this book will help you to accomplish if you are patient and work hard.

What are Limiting Beliefs?

You'll hear the term "limiting beliefs" thrown around fairly often in the personal development world; so, I thought I'd take a minute to define what this term really means.

When you were a child, your inner mind was impressionable, sensitive, receptive and aware. You absorbed the beliefs that others spoke about you in your presence. It could be at a friend's house, from teachers, or from family members. In many cases, you hear beliefs that don't serve your highest self. These beliefs could be:

"Money doesn't grow on trees; you have to work hard."

"Everybody falls in love, but nobody stays in love."

"People are always looking for what they can get out of you."

Beliefs such as these would cause a child to question their innate compulsion to love others without abandon. It may cause a child to work extremely hard in a job they disdain because that's "how it works." In some cases, you were told that you were wrong, or that your innate beliefs were unrealistic, even impossible.

You may have been told:

"You can't be an astronaut; you're terrible at math."

"You'll never be a professional athlete; you aren't tall/strong/healthy enough."

"Everybody in our family dies of heart disease; it's genetic."

In order to become a stronger, more powerful manifester, it is crucial that you identify or name these limiting beliefs. Once you name the limiting beliefs and recognize that you have the power to change them, you will reclaim power over these beliefs. Our words can be blessings or curses. In the case of limiting beliefs, the words you heard acted more as curses. Limiting beliefs are the neurosis, scarcity, fear, or otherwise low vibration of another person passed on to you when you were most impressionable. In order to find the pleasure, prosperity, and love you deserve, you must re-write these beliefs.

It isn't as easy as saying affirmations to reconstruct these negative patterns and programs; it takes repetition, self-belief, and patience. In so many ways, rewriting the patterns, neurological programs, and beliefs with which we were born is the path to enlightenment. Thanks to our understanding of the brain and neuroplasticity, we also understand that it is very possible to rewrite these beliefs to become more free, abundant, and prosperous.

What if you took the six limiting beliefs I stated above and now played the following messages in your head on autopilot?

"I trust my ability to accomplish anything I want."

"I don't have to work hard to enjoy a prosperous life."

"My soul family loves me, embraces me, and looks forward to welcoming me into their world."

"Love does last. Love heals. Love is always possible for me, in all circumstances."

What would happen if these non-limiting, abundant thoughts were on autopilot or embedded into your subconscious network? You would make more decisions from abundant, limitless thoughts. Your business, love life, financial situation, and life in general could become that of what you once only dreamed about.

CHAPTER SEVEN:
Do We "Create" The Bad Things in Our Lives?

"Pain is inevitable. Suffering is optional."
THE DALAI LAMA

We touched on this subject in the introduction to the book, but now we're diving deeper into this question of whether or not we manifest even the "bad" things in our lives. Simply put, we have the power to choose every decision we make starting with our thoughts. However, pain is inevitable. I have heard many strong teachers say that we choose absolutely everything in our lives.

While I consider myself a strong teacher and a decisive person, I am not willing to make a definite call on this one at this time. I don't use the word "everything" very often. And when I do, it's for powerful manifestation mantras such as, "everything always works out for me." However, to say that we choose everything that happens to us is a slippery slope.

The word *everything* is among so many hyperbolic words we use while exaggerating. But more philosophically, words such as "never," "always," "everything," and "everyone" are dangerous because they're so extreme that they often make us liars. When you say that you would never or always do something, you are not allowing yourself room for growth to one day do or not do that thing.

I once had a life coach who encouraged me to be more impeccable with my words by avoiding extreme language such as "always," "never," "everything" and "everyone." What he noticed is that my hyperbolic language made me a dangerous, reckless communicator.

So, I too, struggle to say that something is "always" a certain way unless they're significant scientific basis or spiritual backing. Fundamentally, it's easy to want to compartmentalize a frame of thought or theory by saying it "always" or "never" stands to be a certain way.

* I don't believe we choose everything in our lives.
* I believe we choose damn near everything.
* I believe we are "always" in control of our response.
* I believe that you cultivate the ability to respond in a higher vibrational way as your manifestation practice improves and you make conscious mind-rewiring a part of your daily practice.
* I believe in randomness.
* I believe in being a victim (different from the victim mentality).
* In fact, I believe that we can absolutely have things happen to us that we don't "deserve."

While we don't choose the bad things, we are "always" in control of how we respond, even though some responses are coming from our subconscious mind and programs we downloaded as young as one year of age. I believe that we are, indeed, in control of whether or not we suffer. However, I believe that a life free of suffering isn't an instant choice. It's a daily practice you cultivate over years, or even a lifetime.

One of my favorite phrases from my yoga teacher Bel is this: practice doesn't make perfect. Perfect practice makes perfect. I believe that I'm a better manifester every day. But some days I appear to be the world's worst manifester. Did you like my regression to hyperbolic language? I need to call that life coach again.

Did I choose my divorce? I believe that I absolutely did. I suffered from deep insecurity and codependency that led me to choose a mate that is a typical choice of a codependent. For the record, I didn't know what a codependent was until I was getting a divorce. My friend Jess handed me a book called *Codependent No More* and said that this is why I would always have bad relationships, and how to rewrite that story. At first, I felt ashamed that I'd made such a terrible choice of mates. I also felt compassion and excitement; now that I knew how I got to be in that position, I could prevent it in the future.

Did I choose sexual assault? Certainly not. For one reason or another, I chose the very steps, restaurant location, time of dinner, and any vibration that allowed me to get assaulted. Perhaps it was the push I needed to move (I sought to move within weeks of the assault.) But I was still a victim of this assault. I did not consciously choose to be assaulted. Did I subconsciously put myself in a precarious situation? That's more likely. But the semantics of whether we choose "everything" isn't part of this book. Where we will live is the present moment. I was a victim of sexual assault, absolutely. I may have even subconsciously opened myself up to that attack. And, too, I choose not to suffer any more with guilt, shame, or living in the past moment or trauma of that assault.

CHAPTER EIGHT:
What is Disease?

"Your subconscious beliefs are working either for you or against you, but the truth is that you are not controlling your life, because your subconscious mind supersedes all conscious control. So, when you are trying to heal from a conscious level—citing affirmations and telling yourself you're healthy—there may be an invisible subconscious program that's sabotaging you."

DR. BRUCE LIPTON

A friend of mind was recently diagnosed with cancer, and I was devastated. She stated, "This runs in my family, I expected this." I couldn't help but to ask myself, "Did she manifest this?"

A popular science that corresponds with neuroplasticity is epigenetics. This is the belief that we can alter our genetic disposition, or the concept in biology "relating to or arising from non-genetic influences on gene expression [Oxford]." Scientists such as Dr. Lipton, epigeneticist and pioneer in the research of stem cells at Stanford University, believe that our genes do not have to determine our lives. In the 1960's, Dr. Lipton worked at the University of Wisconsin teaching students that we are products of our genetic "codes." He has done a total 180 in these beliefs since working in laboratories, learning something completely different.

Dr. Lipton discovered that we may not be genetically predisposed to 99% of those things which we think are "genetic". They are manifestations that occur based on the environment. Our environments, including thoughts, are the primary cause of our physical manifestations, with very few exceptions. For instance, one gene can manifest as a cancer in one environment or muscle in another environment.

A concise overview of how this works is this:

1. An environmental stimulus binds to a cell membrane.
2. A chemical reaction inside the cell reaches the nucleus.
3. A gene becomes expressed as a protein.

Lipton proposes that genetics are not predetermined, which is a stark contrast to the teachings most of us heard in school. He confirms the belief of both spiritualists and neuroscientists that our lives are the results of subconscious programs that we "inherit," but that with mental intention, we can actually change this genetic "inheritance." The concept that we can change our genetics based on the environment is the science I referred to a moment ago of epigenetics. Genetics are largely dynamic and pliable; not predetermined and permanent (In a few rare cases, such as certain birth defects, this maxim does not apply).

In order to change our brains through neuroplasticity and manifestation, we must get out of our comfort zones; we must interrupt the patterns we're stuck in. According to Lipton and others researching epigenetics, changing your thoughts won't just increase happiness, but it may greatly ameliorate and even extend your entire life.

In short, through his book *Biology of Belief,* Dr. Lipton illustrates that genes do not control biology, but that our genes receive signals from our environments. Rather than the genes being in control, they submit, rather, to the directions they receive from environmental stimuli. Previous science suggested that genes control cells; positing that we are all victims of what we have inherited from our parents. In this theory, the genes control the cells in our bodies.

Many notable teachers believe that emotional trauma precedes physical manifestation of that trauma. For instance, you can associate a testicular hernia that a student of mine recently cured with his overbearing ex-wife and feelings of not being manly, even being stifled. I manifested my broken toe and my slipped vertebrae during a time of family trauma. When I feel powerless or out of control, my stomach aches.

Many of our physical maladies or disease in the organs comes from repression in the body's primary energy centers, known as the chakras. Neck tightness, soreness, aches in the back, and, yes, even diseases such as heart disease can overwhelmingly be traced back to an acute trauma in a related chakra. Different chakras correspond to different organs. It's uncanny when you become more somatically aware; at once you are more aware of maladies, but you also become more in tune to ways to prevent aches from becoming full-blown illnesses.

> "The function of the mind is to create coherence between our beliefs and the reality we experience. What that means is that your mind will adjust the body's biology and behavior to fit with your beliefs. If you've been told you'll die in six months and your mind believes it, you most likely will die in six months. That's called the nocebo effect, the result of a negative thought, which is the opposite of the placebo effect, where healing is mediated by a positive thought."
> Dr. Bruce Lipton

I'm not a doctor and I'm going to tread lightly on this subject by simply suggesting that if you have manifested any amount of pain or stress in your body, that you explore emotional trauma in conjunction with your current medical treatments. Although I enjoy sharing information and inspiration regarding how miraculously I've healed many of my own diseases through emotional healing, that doesn't in any way mean I expect you to bypass medical intervention in case of an emergency. What I can tell you is that a life of manifestation; more peace, effortless ease and flow have tremendous effects on igniting your body's self-healing mechanisms. Without a doubt, living with manifestation habits reduces stress, anxiety, tension, or fear. Living with this level of abundance helps ignite our bodies' parasympathetic nervous systems, thereby giving our organs more capacity to function properly to help us rest, digest, and heal.

Have you found that your body is weakened during or following a time of stress or emotional trauma? Let us know more about your story at www.facebook.com/groups/manifestingsecrets/

Fight, Flight, Freeze, or Fawn

Another critical term to understand to manifest your dream life is the difference between the sympathetic and parasympathetic nervous systems. We've touched briefly on these concepts, but now we'll look at how they play into manifestation.

Regulation and optimal functioning of your parasympathetic nervous state is becoming more a part of the collective consciousness than ever in recent years as teachers and healers in the holistic space have exploded; providing greater depth of information and care than a primary physician with whom you may only spend a few minutes a year. Simply put, we're becoming more curious about our bodies, and we're enlisting healers who invest time in educating us rather than prescribing pills and showing us the door. Gone are the days when people suspected that a McDonald's hamburger was healthy because it had "protein" or that high-fat diets necessarily make you fat. We've evolved into so much awareness of nutrition, preventative medicine, and the role of emotions in disease.

That role is inarguably a starring role, if not the main actor.

You may recall that a healer I visited encouraged me to "drop, drop, drop" when she worked on healing my back. She was telling the body to relax; to transition to a state of ease, rest, and digest. In your sympathetic nervous state, or SNS, you're in what's commonly referred to as fear, or "fight, flight, freeze, or fawn." When we fight, flight (abandon a situation), freeze (experience paralysis from our fear) or fawn (placate an abuser or threat to stay safe), our brains can't manifest a new reality. Our sympathetic nervous state shuts down most of our non-vital functions, including the ability to remap negative thought patterns peacefully and consciously. In many ways, this is a book on how to live a life of peace, stillness, surrender, and happiness.

Scientifically, we are unable to evolve while in our sympathetic nervous states. Neuroplasticity, healing, learning, and growing are impossible while we are in "fight, flight, freeze or fawn." It is not realistic to completely rid ourselves of sympathetic nervous states; it's a useful mechanism designed to protect us. It gives us the power to escape dangerous situations or people. The fight, flight, freeze, or fawn instincts are protective mechanisms designed to keep our primal ancestors from getting eaten, basically.

The problem is that our brains build maps based on traumatic events. Many scientists, including Dr. Joe Dispenza, teach about how we emotionally revert to traumatic events every time we remember something traumatic that happened to us. We may relive the experience of a loss or death or any other traumatic situation in between those two extremes every day - nay - every hour of our lives for years, or even decades after we're otherwise 'safe' from the threat.

In his book *Becoming Supernatural*, Dispenza details the story of a mother of three children who became widowed when her husband committed suicide. She replayed the fear, anxiety, hurt, anger, and rage in her mind over and over for many years. Prior to the suicide, the woman was healthy with a thriving career. After her husband chose to commit suicide, the woman's body failed; she chose to be in a relationship with an abusive lover, and she suffered financial ruin from the inability to work and codependence on her abuser. The traumatic event put her in such a state of dis-ease that her body began to shut down.

It was through many of the steps in this Manifesting Secrets training program that the woman summarily changed the course of her life; she is now healthy, happy, and speaking publicly for scientists such as Dr. Dispenza. The process that saved this woman's health, wealth, and happiness was fundamentally her consistent, daily decision to move from a place of pain and sympathetic nervous states to her parasympathetic nervous state.

Turning on the Parasympathetic Nervous State

I'd be remiss if I didn't share with you some of my personal favorite tools for activating the parasympathetic nervous state. One tool that I use every day is deep breathing. I breathe in and out through my nose, sometimes I even practice this while running up the mountains where I live in Aspen, Colorado.

Another tool I combine with deep breathing is to visualize my breath as I expand my lungs as an act of receiving light and love while focusing on breathing into my heart. As I exhale, I visualize that light filling my home or I direct that light to somebody I love, or somebody with whom I'm angry and want to manifest a conflict resolution.

I further expand this practice by breathing deeply and meditatively with my legs up the wall. This has been transformational in quelling former back pain, and it's even said to have anti-aging benefits.

Other parasympathetic activation techniques I use are meditation and breathwork, more of which I'll discuss in our seven steps in Part Three. My favorite meditations are sometimes guided meditations such as Yoga Nidra or else a variation of Dr. Dawson Church's Eco-Meditation that incorporates EFT Tapping techniques, which I'll discuss further in Part Three.

Activating the vagus nerve is the biological way to support your parasympathetic nervous system and can be done by something as simple as dropping your tongue from the roof of your mouth and breathing deeply. An advanced method for vagus nerve activation is commonly referred to as "gut smashing," whereby the practitioner lies face-down on the floor with a medium-sized ball and rolls around while breathing deeply. It has been hailed by Shawn Stevenson as a method for deeper sleep.

Not to be forgotten is the newly popular Wim Hof breath and ice bath craze. Biohackers and peak performers like Tony Robbins promote daily cold plunges for optimal parasympathetic nervous state functioning. It has almost an instant effect on reducing stress; and it's about the quickest way you can enter the present moment imaginable.

CHAPTER NINE:

Next Level Brain Healing with NuCalm

In early 2022, I was about eighteen months into my legal battles, periodically having doubts about my victory, and going through a blue period of wanting a long Aspen winter to end. I felt generally satisfied with my purpose; I'm not prone to depression and I almost never doubt my purpose. *A strong sense of purpose is the best antidote to depression I know.* However, the horrific legal abuse had heated up and now I wasn't fighting one battle, but two. I hired an attorney to fight in the Appeals Court alongside the District Court battle I was already engaged in. Due to a mistaken text message sent to me by my son's former doctor's office, I learned that my son was scheduled for a mysterious doctor's appointment, an "annual wellness checkup" taking place in the middle of the school year at a doctor I'd already indicated I wasn't comfortable seeing. A doctor who had a record of me saying that I would not allow my son to take vaccines until I was present and had discussed risks and benefits with her and, after having held out for five years, caught my child up on the childhood schedule in one appointment. In a word, what she did was homicidal.

What's more, I had made written requests to the doctors' offices as well as to my son's dad.

I didn't trust this irresponsible doctor, and I had good reason not to. In my opinion, she was a danger to children, recklessly insensitive, complicit in obstructing my authority as the child's mother, and a menace with a syringe full of experimental concoctions that could harm or kill my child while adding no conceivable benefit to his life. She could have made my son a statistic, like so many other sick and dying kids who were subjected to the experimental mRNA Covid shots.

Do you ever catch wind of something, and you don't have evidence yet, but you know beyond any shadow of a doubt that it just doesn't feel right; that something nefarious is happening? At that moment, sitting in my office in Carbondale, Colorado, I knew that something purely evil was happening in collusion with the doctor. My stomach dropped, my head was filled with a sort of droning alarm like a tornado warning, and my entire face felt like it was on fire.

When I called the doctor, even though they had a record of me being his mother and having taken him there since he was an infant (until her vaccination without parental consent in 2020), they wouldn't answer questions.

I entered the office and asked them to confirm the appointment. I then asked what was on the agenda. They acted like they had no clue what I was talking about. I then asked the office to cancel the appointment, to which they responded that they could and would not. I drove back to my office to retrieve legal paperwork indicating that there was an active district court case that dealt with medical custody of my child, and that as the co-conservator of my child's health at that time, they had no authority to, well, skirt my authority. I was verbally assaulted by the staff.

In the meantime, I texted and called my attorneys and their assistants. "What can I do? Did he find a loophole? What is happening? Have we lost? Are they going to potentially kill my son?" I wasn't playing around with this experimental shot.

My attorney determined that they may have incorrectly determined a six month stay that I had from a previous trial, and that they may have had the intention of inoculating my child.

Back at the office, they lied to me and told me the doctor wasn't on site. I quickly caught them in their lie. The doctor came out and, because I had legal documents in my hand, chose to allow me to cancel the appointment. However, she wouldn't tell me that. She played like she didn't have the authority to cancel it. She said she didn't see how the legal paperwork was her business. She claimed that I needed to upload it into their complicated medical system and that she couldn't guarantee how long it would take to process.

I was useful and productive for the following few days but exhausted. That incident, on top of eighteen months of legal abuse, threats of jail, trials, a deposition that bordered on verbal assault and sexual harassment, and contempts of court had nearly broken me. As I'm writing this manuscript, I have invested well over one hundred thousand dollars to protect my son in the past two years. I will never regret it nor look back on that decision; but the legal system is far and away the worst place for a parent to find themselves. The attorneys are heartlessly motivated to inflict pain on the other party in a way that invariably and inevitably damages the child. My son has been privy to information that he doesn't understand, but that caused him to begin having night terrors and exhibiting other behavioral indications that because of the unnecessary legal fighting, he is the one who is most suffering.

In a word, I am a warrior. I'm a soldier in God's army. I'm a relentless truth-seeker. Since the last edition of this book, I've had some tough battles, but I've also handled them with more efficiency, emotional regulation, and elegance than I could have fathomed even during the first edition of this book. In short, the hard things have not hardened me; they've made me a more powerful earth angel; I'm not afraid of hard things; I'm invigorated that God chose me to be a warrior for Truth.

Nevertheless, I was tired. Anybody forced into fight or flight on the battlefield as much as I had been over the previous two years would be.

An author of mine who is a beautiful man and neuroscientist recommended NuCalm several months prior when he saw my acuity and positivity somewhat diminished. I didn't embrace it. Then another author of mine, a former sniper who had suffered several tragedies even since he retired from the British Army, began using NuCalm to treat his Post Traumatic Stress Injury (PTSI). After his first experience with NuCalm, he shared with me that he slept better than he had in years, maybe more than a decade. He was so passionate about the healing power of NuCalm that I was finally sold.

Waiting until Hunter went to his daddy's house, I finally laid down one night with NuCalm. After my first seventy-five-minute session, I sat in my room for hours peacefully staring at the wall in front of me in the pitch dark room. I didn't have any music or noise. I just stared. For the first time in longer than I could remember, my mind was very quiet. I went back in for a second NuCalm session. When I woke up, it was nineteen hours later.

If you've suffered any type of trauma that has caused disruption to your healthy sleep, you know what a good night of shut eye is like. In my experience, it's more powerful than a two-week beach vacation.. It's truly, literally life-altering.

I began using NuCalm regularly for months. I even started having my son use it and his nightmares disappeared almost instantly. I began working more cleverly; within weeks my calendar was full of twice as many authors and many true celebrities and heroes in the medical freedom movement that was so dear to my heart. Most importantly, I felt even more emotional regulation and equanimity in my life. I've always been patient with my son, but sometimes my calm visage was just a mask for severe irritation or internal conflict about his growing behavior issues. I found myself literally smiling when he acted out. I could calmly tell him that Mommy needed this or that from him, and I didn't feel anxiety about his tantrum or irritating behavior. I literally felt like I could handle anything.

I felt empowered, victorious, and limitless, even beyond my previously awesome state of mind and Be-ing.

What Happened to Me on NuCalm?

Your brain is complex, and in this book, we'll learn about how you can train your brain to materialize your true essence and desires. Your brain has millions of cells forming trillions of connections using neurons. Neurons communicate through chemical and electrical signals, creating pulses known as brain waves. A neuroscience technology that has been patented and clinically proven is called NuCalm, and it allows you to lower stress, fuel recovery, increase concentration, and manage a variety of challenges your day may bring. Feeling stress or anxiety,worried, or overwhelmed? NuCalm helps flips the switch on stress by downregulating your sympathetic nervous system, another concept you'll encounter in later chapters of this book.

To enjoy NuCalm, simply find a comfortable place to relax, put on a quality headphone, select the outcome you want (channel) and press play. Neuroscience does all the work for you, so all you have to do is take a deep breath, relax, and enjoy the journey. Now, how does NuCalm work? Allow me to try to simplify complex neuroscience. NuCalm incorporates physics, mathematics, and algorithms inside of a neuroacoustic software that lies underneath music. This software, when listened to through quality headphones, presents your brain with a signal that your brain follows. For example, NuCalm Rescue guides brain waves from high frequencies (stress / anxiety) to lower frequencies (alpha & theta) that are associated with relaxation, recovery, and restoration. This complex software is easily accessed through a mobile app; it's like having a remote control for your brain in the palm of your hand. Simply select the channel you want and press play. If you want to focus, listen to Focus; if you want to recover, listen to Rescue; if you want to feel creative, listen to FlowState; if you need a quick recovery and healthy energy boost, listen to PowerNap; if you want to perform at peak levels with high intensity, listen to Ignite; and if you want to sleep, listen to DeepSleep.

NuCalm used to be an expensive FDA Class III medical device exclusively available to the US military, pilots, doctors, professional athletes, and cancer patients. The most impressive data point for this technology is the fact that NuCalm has been used in over 2,000,000 surgical procedures replacing general anesthesia. But in November 2021 the company successfully transformed the NuCalm medical device to a consumer subscription available through a mobile app. The new NuCalm is easier to use, more affordable, and provides greater consumer choices, from the deepest levels of sleep to the highest levels of intensity.

In addition to the powerful recovery and restoration of Rescue, I love to listen to Ignite, which increases brain waves to Gamma (39Hz - 41Hz), which helps fuel your cognitive function. I listen to Ignite™ while working or when I'm feeling sleepy but don't have time nor desire to nap or use caffeine. The Focus™ journeys fuel Beta brain waves to help to enhance your focus, concentration, and comprehension. Focus oscillates brain waves between 15 Hz - 20 Hz, which is the sweet spot for learning. I use

Focus when writing books or working on graphic design in my business. Rescue™ is my "go to" for daily restoration and I find it effectively puts my brain into a healthy state and has allowed me to process and heal trauma without drugs. That's a big deal for me. The Deep Sleep™ journeys gently guide your brain waves into Delta and, alas, help you get remarkable, deep sleep. My son loves Rescue™ "40", and I enjoy the Rescue "30" or "50" minute journeys.

Whether you're healing from acute trauma that you can name or looking to heal your brain from trauma that you can't fully recollect, NuCalm is a technology you can use to help your brain function in the way you need it to for the task at hand... or in the case of Deep Sleep™, for no task except restorative, peaceful sleep.

If neuroscience intrigues you and you want to learn more about NuCalm, in 2016, *A New Calm* was published. This book, authored by the esteemed Dr. Michael Galitzer and Larry Trivieri, goes deep into the neuroscience of NuCalm and the etiology and biology of addiction. Also, the NuCalm website (www.nucalm.com) has a lot of information as the company has been educating doctors for more than 14 years.

CHAPTER TEN:
Our Thoughts Create Our Realities

"If you realized how powerful your thoughts are, you would never think a negative thought."
CAROLINE LEAF

Our thoughts create our realities. Our thoughts create the good and the bad manifestations in our lives. We are always manifesting. Our goal, then, is to become more conscious manifesters versus subconscious "robots" playing out old subconscious programs. Although we can't always control the people and events around us, we always have full authority in our minds for our response to these events and whether we're going to make an opportunity out of obstacles in our paths.

My favorite analogy of life before and after Manifesting Secrets comes from the movie *The Matrix*. In the story, the protagonist Neo is given the choice to wake up and see the truth of what the world really is, or to be naïve and asleep, allowing his limited views and subconscious programs to guide his life. In some ways, it's also a metaphor similar to the *Allegory of The Cave* written thousands of years ago by the philosopher Plato. Today, this analogy has become common vernacular among people who consider themselves "free" from the matrix. The modern day matrix is, among other things, these eight pillars of wealth and power control:

1. Big Tech
2. The Fed/Finance Cabal

3. Big Academia
4. Media Manipulation
5. Big Government
6. Poisoned Food Supply
7. The War Machine
8. Big Government

A tool these pillars of manipulation and evil use is fear. What's more, if you dig deep, they use children as currency.

Another story: Plato's The Allegory of the Cave: A man is in a cave viewing the world by its shadows on a wall. He sees a shadow of a rabbit and calls that a rabbit. He only sees shadows of what is "real" because of his limited point of view from the cave. This is how we live before we "wake up" or become conscious.

Experts believe that up to 95% of our lives are the result of subconscious programs we downloaded before the age of 7. The way we think, act, feel, believe, worship, respond, communicate, accumulate wealth, and even heal or die all come from these subconscious programs or thought patterns.

Our subconscious or "below conscious" programs have actual physical structures in our brains. What we see and experience is filtered through the lens of those programs; we aren't making conscious choices but, rather, relying on automatic programs developed by caregivers, parents, society, media, trauma, and especially from emotional highs or lows. Dr. Bruce Lipton likes to say that The Matrix isn't science fiction—it's real life.

Because you have chosen to pick up this book, you have chosen the "red pill." You want to see more than the shadows. You want to dive deep below your subconscious mind and into the programs, beliefs, behaviors, habits, and patterns to consciously choose the way you live, love, accumulate wealth, and feel in your mind and body.

Scientists believe we think over 60,000 thoughts daily, and the subconscious mind processes an astounding amount of information without needing you to consciously make the choice to think about those things.

CONSCIOUS & SUBCONSCIOUS MIND

Conscious Mind	Subconscious Mind
LOGICAL	EMOTIONAL
Newer	Older (lizard brain)
Weaker	Most Dominant
Tires Easily	Always on
Not Automatic	Automatic
40	**40,000,000**
bits of data per second	bits of data per second

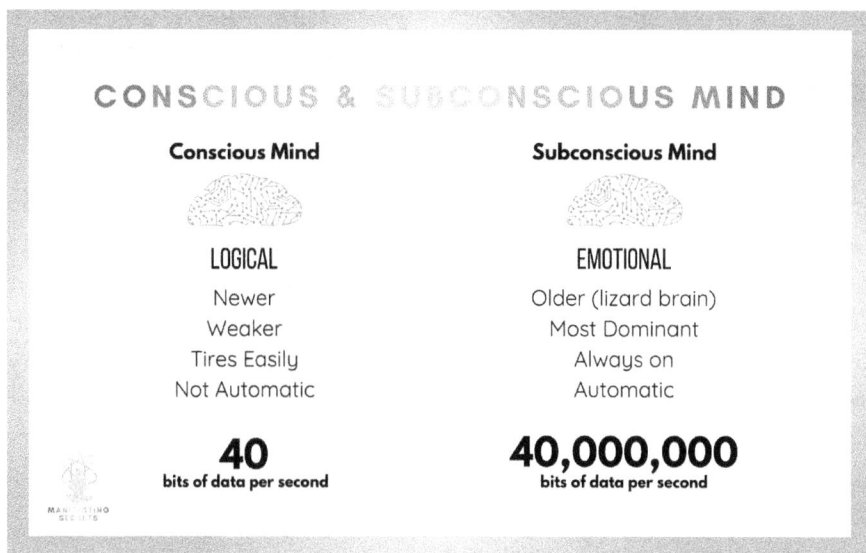

Negative thoughts attract negative emotions and become negative material manifestations.

Positive thoughts attract positive emotions and, therefore, attract positive manifestations such as "wealth, health, love, and happiness." This concept is meticulously detailed in popular Law of Attraction literature. However, you will learn that although you may diligently force yourself to think positively, you will not experience positive manifestations until you remap your brain or bring more positivity to the subconscious level and create the habits that are needed to make that positivity a way of life. The abundance compounds like a snowball or even an interest-bearing bank account. You begin with positive thinking, which helps motivate you to create manifestation habits. But each day you deposit more manifestation habits into the bank account. The growth compounds and elicits greater rewards over time.

The goal in creating powerful manifestation habits is to gradually remap your brain. When you physically alter neurological pathways, you transmute negativity into positivity. It works like a computer. Your brain is the hardware; your thoughts are the software. You can install new software through manifestation habits. This helps you to gradually delete the negative thoughts and hardwire more positive thoughts. This, subsequently, hardwired a life of more health, love, peace, and wealth.

While rather straightforward, the process of remapping neurological pathways isn't altogether easy. Remapping the brain to trash negative thoughts that sabotage your health, relationships, wealth, and happiness requires consistent, daily, repetitive effort and unending patience.

Remapping old patterns doesn't happen overnight. Sadly, this sobering fact isn't as well-popularized in positive thinking literature that encourages you to say affirmations and expect cars to show up in your driveway. It sells a lot of books, but it doesn't remap many brains. If anything, such overly optimistic and simplistic books on manifestation have led many people to feel ashamed of their "bad manifesting" efforts or even defeated; unwilling to give manifestation another try for their misunderstanding that manifestation should happen overnight.

CHAPTER ELEVEN:
Manifestation is Possible for Anybody

"Your beliefs become your thoughts,
Your thoughts become your words,
Your words become your actions,
Your actions become your habits,
Your habits become your values,
Your values become your destiny."
GANDHI

Manifestation is possible for anybody who can think. You may be struggling with health issues or poverty and you may have a long journey ahead of you. It's important to know that a life of manifestation starts out slow and compounds as you live with daily, consistent manifestation habits. My favorite teacher from middle school French class, Ms. Nelson once said, "You can eat the elephant if you do it one bite at a time."

In the book *Atomic Habits* by James Clear, we learn that an improvement of just 1% daily results in an astounding 3700% improvement by the end of one calendar year. Imagine if your marriage improved so tremendously in just one year because of your daily efforts and diligence? What if your finances improved so dramatically because of daily, consistent practice, education, and confident choices around money? Would starting out small be worth it to you if you could see the bigger picture?

JAMES CLEAR'S COMPOUND INTEREST OF SELF-IMPROVEMENT

If you improve just 1% each day for a year, you'll have improved over 37 TIMES by December 31st.

What would an improvement of 3700% mean in your health, wealth, or love relationships?

COMPOUND INTEREST FOR SELF-IMPROVEMENT

1% Better → 1.01˙365 → 37.78

1% Worse → 0.99˙365 → 0.03

1 YEAR

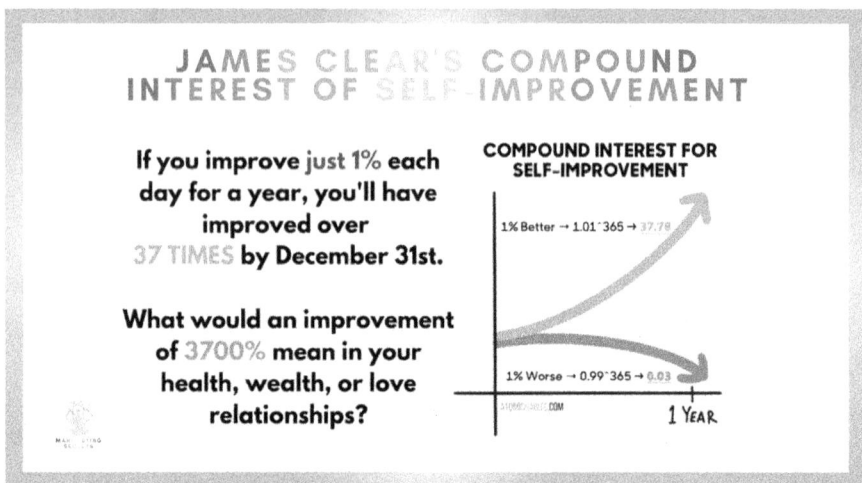

You're absolutely qualified to manifest things your parents or grandparents never dreamed of manifested in their lifetimes. Critically, that requires daily habits that help change the thought patterns, beliefs, programs, and behaviors that you learned as a child.

Your mind is capable of manifesting even if you can't move. Not only have I had the chance to watch the Neurofunctional Institute's work on health quadriplegics, but great teachers like Dr. Joe Dispenza tell stories of having healed from paralyzing injuries in their minds alone. My friends Michael Warner (author of *Holy Cow*) and Ron Michael (founder of Ascension Keys) used their minds to walk again after doctors told them it was unlikely they'd live after their respective car accidents, much less walk again. Manifestation starts in the mind through mental intention.

Pascal's Wager of Manifestation

If you've studied philosophy, you've likely heard of "Pascal's Wager." It's the concept that you never lose when you believe in God. As a manifester, you are being called to believe in yourself, even if you don't see the material manifestations come to pass right away. The day that I laid on my back in my living room and meditated for four hours that the pain in my back and foot would cease was the day I understood how important the power of self-belief is. One thing is guaranteed: if you don't believe in yourself, whether or not you have the ability to heal your brain or body, you never will.

These charts will help. In the first one, you see the traditional Pascal's Wager. If you believe in God, you experience peace in the promise of Heaven even if He doesn't exist, and if He does exist; you experience the reward of eternal happiness. If you don't believe in God and He doesn't exist, nothing happens (but you may live a life with less security and peace). If He does exist but you don't believe, the Wager states that you will lose out on the potential eternal reward of communion with Him.

PASCAL'S WAGER

	God Exists	God Does Not Exist
You Believe in God	Eternal Happiness (= Heaven)	Nothing Happens
You Don't Believe in God	Eternal Damnation (= Hell)	Nothing Happens

MANIFESTING SECRETS

I use a similar chart for manifestation. Believing in yourself or believing in manifestation is like belief in God. You have nothing to lose by believing in yourself and your ability to create endless love, health, wealth, and happiness in your life. If you don't believe, your negativity guarantees your failure at manifesting. If you do believe, you're living with hope, confident action, and positivity. Believing in yourself also helps to attract good things into your life through the Law of Attraction.

THE PASCAL'S WAGER OF MANIFESTATION AND THE LAW OF ATTRACTION

	Manifestation is Real	Manifestation is Not Real
"I Believe I Can Manifest"	"I love better, heal faster, and experience in abundance."	"I Live a Life of Hope, Positivity, Healthy Habits and Love."
"I Don't Believe in Manifestation"	"I Miss Out on the Joy of Remapping of My Mind for Love, Health, Wealth and Happiness."	"My Negativity Persists and Perpetuates More Negativity."

CHAPTER TWELVE:
Where Spirit Meets Science

"The most beautiful and most profound emotion we can experience is the sensation of the mystical. It is the sower of all true science. So to whom this emotion is a stranger, who can no longer wonder and stand rapt in awe, is as good as dead. To know that which is impenetrable to us really exists, manifesting itself as the highest wisdom and the most radiant beauty which our dull faculties can comprehend only in their primitive forms-this knowledge, this feeling is at the center of true religiousness"
ALBERT EINSTEIN

The mystical and scientific aspects of manifestation are not opposing; they work in conjunction with one another. What makes you a powerful "manifestor" is uniting the metaphysical with the material through reverence for the science and mechanical habits that tether one to the other.

The science with the "wu-wu," the yang with the yin, the positive thinking as it relates to neurology; the most progressive studies in science corroborate the beliefs of ancient mystics, seers, and saints. As you declare what you want with intention, you will find that not only will the universe shift to align to your highest state, but you create the physical structures of the brain to turn thoughts into reality.

Your definition of miracles may change as you watch the seemingly impossible become possible through our expert teachers and research in this book and in your Manifesting Secrets Brain Training program.

Neuroplasticity 101

Neuroplasticity is the science of changing the brain through mental intention. We've provided a quick definition of neuroplasticity in preceding chapters, but now we'll go into more detail.

Neuroplasticity is a big scientific word we'll use over and over again in this book, and I hope you will use it in your life as well. Neuroplasticity includes the root "neuro," which means brain. If you're like me, the concept of understanding the brain is in and of itself rather intimidating.

So, what is neuroplasticity? Is it a fad? A "buzzword?" Is it a farce? Neuroplasticity is a science that has been exploding in mainstream popularity in recent decades.

However, it's not a new science. It's as old as the human brain. In fact, the world's most revered ancient texts (even many that pre-date the Old Testament), are full of wisdom about neuroplasticity that were given through inspiration and intuition through sages throughout the millennia. Nearly 2,600 years ago, the Buddha said:

> *"What we think, we become.*
> *What we feel, we attract.*
> *What we imagine, we create."*
> BUDDHA

Our thoughts become words, which become manifestations, which become material reality.

I have been a student of creating a reality different from the one I was programmed to live in childhood for at least ten years, perhaps twenty. Have you ever felt like you didn't belong in your school or family? Or like you could see dimensions above and beyond the surface reality you were living? This is a good indicator that you've been awake or aware of your conscious or even subconscious programs from that moment.

The programs we inherited from childhood are so ingrained in us that we don't even know we're operating on them. The way you behave, speak, respond, react, communicate, move, eat, and almost anything you can conceive of isn't entirely a function of free will or free thinking.

Perhaps you've heard of neuroplasticity or brain plasticity. You may have heard that your brain is malleable, and that you can change your life by changing your brain. And if you're like me, that sounded like a lot of research, information, details, and even science that is likely above my head. Happily, the most important things you need to understand about neuroplasticity are remarkably digestible, and even simple.

Your brain has a natural ability to restructure its neural networks. This allows nerve cells to adjust their formation based on your environment, your stress or anxiety levels (e.g., fight or flight), or other critical events, including physical injury. Such neural activity happens automatically; your mind is constantly firing and wiring in conjunction with your thoughts and environment. More critically, your mind is already firing based off of past events, thoughts, and environments to which you've been exposed.

Many of our current neurological programs aren't serving us. Perhaps we interacted with an environment, person, or belief system that caused us to experience fight, flight, freeze, or fawn. Due to these events, neural programs that are wired together protect us from future trauma.

Your neural programs develop in two unique ways.

1. Gradually. Many of your subconscious programs were developed before the age of seven and were formed by watching and listening to your caretakers.

2. Instantly. Some programs were developed instantly while in highly emotional states, painful events, or trauma. A great example is a car accident. Many people who experience scary car crashes will take years to return to the same feeling of safety and security on the road. Their brains are wired to associate cars as dangerous

because our brains have one goal in mind at all times: to protect us. Your brain doesn't seek to make you happy. It seeks survival. Sometimes we replay traumatic events in our minds over and over, even decades after that painful event.

Scientists believe that by age 35, 95% of your life is the result of subconscious, hardwired programming in your brain. This means that we aren't living our lives based off of free will. If we are, indeed, the result of subconscious programming, then we are merely puppets to our subconscious minds. We are slaves to the old programs.

Do you believe that certain bad choices or experiences in your life were not your "best self," but choices you made due to your subconscious programming? It gives you pause; and it hopefully fills you with a sense of self-love, self-compassion, and grace for yourself. Do you also believe that you are powerful enough to rewire the negative programs to create a more positive life? Of course you do. That's why you're here. We're here to rewire the sadness, pain, trauma, depression, and subconscious negativity in order to create happier, more peaceful, more stable, more prosperous, and more consciously controlled lives.

Two of the most important principles from neuroplasticity are these:

1. Neurons that fire together, wire together.
2. If you don't use it, you lose it.

When you understand these two principles, you've already unlocked one of the most important secrets to manifestation.

Neurons That Fire Together, Wire Together

With the power of your mind through neuroplasticity, you can alter the synaptic connections that create your thoughts, behaviors, beliefs, and other programs that manifest your material reality. Synaptic connections in your brain are constantly altered by the neurons that carry them. Those connections are moved, removed, or recreated summarily in a process called synaptic pruning. Just like pruning in gardening, the brain eliminates connections that don't get much use.

Your brain takes in up to 20 million billion bits of information every second. That's not a typo: that's a lot of information. As you experience the world, you make synaptic connections. Synapses that are consistently used have stronger connections. This is where the phrase, "neurons that fire together, wire together" comes from.

On the other hand, synapses that fire infrequently will be eliminated, i.e., "use it or lose it." These connections are designed to give us a brain circuitry that preserves our survival in the world in which we live. Synaptic pruning begins just after birth, when babies' brains dissolve connections that are no longer relevant for survival after being in the womb, and the pruning continues into our adult lives.

Growing up in a small family, growing up rich or poor, having parents who communicate well versus those who fight often; these are all behaviors that affect our overall neural circuitry through synaptic connections. We form connections based off of the programs that are running frequently. If we don't use a program much, it is dissolved.

Your brain is not inclined, however, to prune negative programs. It doesn't judge negative programs or synaptic connections. For instance, if alcohol abuse, smoking, a raging temper, and porn addiction got you this far, and you're alive, the brain doesn't seek to change that behavior or program. Your brain, once again, is focused on one thing: survival. If you're alive and avoiding what the brain perceives to be a threat, then the brain is happy.

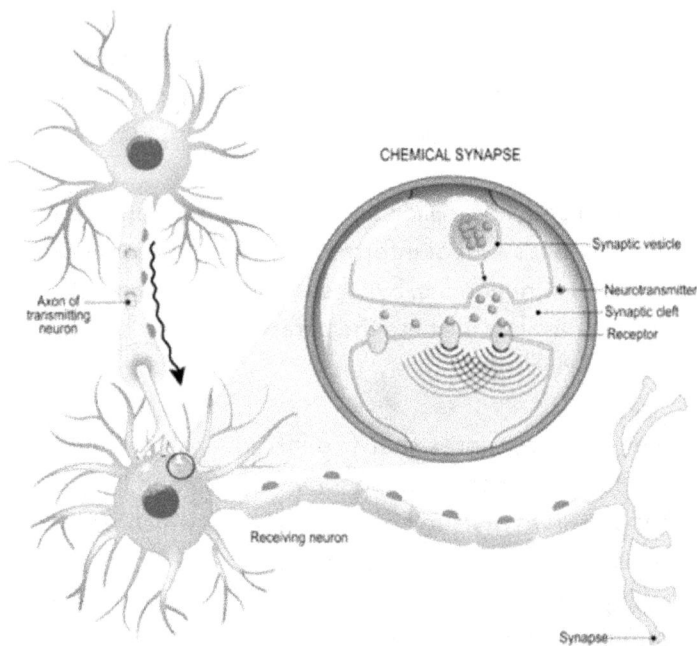

NEURON COMMUNICATION

CHEMICAL SYNAPSE

Axon of transmitting neuron

Synaptic vesicle
Neurotransmitter
Synaptic cleft
Receptor

Receiving neuron

Synapse

Storing space, so to speak, for synaptic connections you don't use is exhausting for your brain. Although it's a marvel and we will likely continue to study the brain for all of human existence, the brain nevertheless has limited capacity. That's why if you don't use something, it is eliminated. While only comprising 2% of your body weight, the brain consumes 25% of your calorie intake. In babies and children, the brain is believed to use between 47-88% of the energy consumed by the body while at rest. Researchers believe that this is why human adolescents spend so much more time in childhood. When the brain consumes this much energy, body growth slows. Some mammals reach adulthood in a fraction of the time as humans. It's magnificently busy.

Did you know that when you look at pornography, your brain doesn't actually "realize" that you aren't engaging in the act you're viewing?

Did you know that when you visualize something painful disappearing, you begin to feel the release of tension over that situation?

Did you know that you can become more empowered just by visualizing your own power over something?

Your brain, alas, doesn't know the difference between what it sees materially versus what it perceives in your mind's eye. Your brain, just like driving, steers in the direction of what you focus on. Why is this? Simply put, your brain doesn't differentiate between your mind's eye and the "material" world because the same specific neural networks fire during both occasions. In our brains, nerve cells create maps based off of what "moves" us; based off of what we focus on and based off of what triggers high emotional states. When we focus our thoughts on an event, an image, or a situation, whether good or bad, our brains create programs to respond to that event, image, or situation. Hence the "wiring together."

If you grew up in an abusive home, your brain may have created a network of programs to protect you from abuse. This may sabotage your emotional health by making fight, flight, freeze, or fawn a normal state for you as opposed to an emergency protection mechanism. If you experienced a death in your family with which you haven't made peace, you may have developed programs that prevent you from enjoying future relationships that could hurt you again.

Understanding this concept of developing or pruning synaptic connections is critical to manifestation because it shows us that the brain is malleable and changeable. Therefore, so too is your life. You aren't "stuck" being poor, overweight, dependent on stimulants or drugs, or even depressed. Your thoughts can bring freedom from despair in the same way that it has, up until this point, potentially causing you tremendous pain.

Rewiring Through Repetition

Each day, we wire our neural circuitry through our habits, rituals, and behaviors. Oftentimes, I feel compassion for students who approach me and ask, "Stephanie, teach me how to manifest a job, rent money, and the love of my life right now."

While I have a plethora of exercises that can assist with calling in love, health, or wealth, manifestation is a marathon, not a sprint. Although a 72-hour manifestation retreat can absolutely alter the course of one's life and bring them never-before-experienced peace and satisfaction... the truth is that when you return home from such an experience, you still have the responsibility of creating the permanent synaptic connections to support the "new you." This requires a couple critical things:

1. Creating a life of permanent, habitual manifestation through wiring the good things together and pruning out the bad things requires accountability. Your brain, once again, isn't interested in your happiness and can be very resistant to change, even change that you know is good for you. My students spent at least 90 days performing exercises in Manifesting Secrets that can sometimes feel repetitive, such as daily guided affirmations, gentle neuroplastic movements, guided and self-directed meditations, and journaling. However, these habits are homed in during your first 90 days of brain-training so that you will make the fruit of these habits a new way of life. That fruit includes, but is not limited to, deeper peace, connection with God, confidence, a healthier body, a clearer mind, better sleep, and a resounding sense of optimism, faith, and hope for the new future you are creating

2. The second critical component to making manifestation a lifelong habit to create an ever-better life is repetition. If you say one single affirmation 10 times daily for the next 90 days, you are far better off than if you had focused on 1 million unique affirmations over the same period. That's because repetition is what creates those grooves, pathways, programs, and synaptic connections that physiologically alter your mind. Remember: neurons that fire together, wire together. One act repeated over and over will eventually become more effortless and automatic with time, such

as a free throw shot. However, if you practice 20 different new activities over the same period as you'd have practiced your free throw, you'll likely remain a novice at all of those activities.

Repetition is critical to your success in neuroplasticity; and it burdens my heart to see people market and profit off the promise of instant manifestation. That's just not how the brain works. As I like to tell my students: your brain doesn't rewire based on one-night stands, but long-term relationships with your habits.

For instance, if you feel anger and replay the moment you found out your boyfriend cheated on you every day, you will rewire and reintegrate that neural network according to how much attention you devote to it.

If you think thoughts of love, compassion, and goodwill, meditating or praying for others, you'll rewire your spirit and relationships accordingly.

If you play victim in your life and blame others for your pain, you will manifest a feeling of being violated. Being violated will become your default feeling, and you'll attract violation in your life.

The good news is that even if you're experiencing:

Fear
Anxiety
Dis-ease
Loneliness
Lack of love

... you can remap those habitual thoughts with dedication to repetition of new thoughts and habits.

You don't have to feel fear.
You don't have to feel anxiety.
You don't have to feel dis-ease.
You don't have to feel lonely, ugly or self-conscious.
You don't have to feel a lack of love.

With neuroplasticity, you have just found the keys to true freedom.

CHAPTER THIRTEEN:
Interrupting Subconscious Patterns to Manifest

"We can't solve problems by using the same kind of thinking we used when we created them."
ALBERT EINSTEIN

I first heard of the phrase "pattern interrupt" when my friend Jon used it to teach marketers how to get people's attention by writing or posting something that they would never have expected to see. It has since permeated the vocabulary of any marketer worth their salt. There's a lot of psychology behind the concept. You see, most of everything you do comes from a subconscious pattern, most of which, as we discussed, were developed before the age of seven. How we interact in relationships, how we think about money or health; most of these inclinations are a result of subconscious or "below conscious" patterns. We're interacting with the world and even within our own minds on autopilot. The way things have always been done are the ways we will continue to think and act.

When we interrupt our thought processes and patterns, this produces a chemical response in the body. Each time we interrupt the nerve cells that are connected to each other, we break up their long-term relationship just a bit more. Albert Einstein brilliantly said, "We can't solve problems by using the same kind of thinking we used when creating them."

As you go about your day, you reinforce either your positive or negative habits with physiological responses. For instance, MRIs of the brain indicate that angry thoughts send a surge of blood to the right prefrontal cortex, associated with depression and anger, but happy thoughts send that blood surge to the left prefrontal cortex. MRIs have revealed to us that we reinforce what we think about via that blood flow. In other words, when your blood surges to either the left or right prefrontal cortex, you refuel that feeling. Knowing this, why wouldn't we continually send our brains happy thoughts to regenerate more positivity and happiness?

What You Think About Manifests

Remember that you are always manifesting, whether subconsciously or consciously. Our goal in Manifesting Secrets is to move from robotically following subconscious patterns to consciously rewiring what we want into our brains. I believe that love, health, wealth, and peace can be your default programs over time with persistent, patient manifestation habit-forming.

When we interrupt the thoughts that produce chemical responses in our bodies, we dissolve that hardwired relationship or wiring. Cells that don't fire together are no longer wired together. Neuroplasticity is, therefore, your way of taking back the power over your life.

Shortly after I became a mom, I learned from Dr. Lipton that children during their first seven years of life are biologically set in a theta brainwave state in his book *The Biology of Belief*. It compelled me to study this concept with passion. Before scientists had the technology to corroborate it, Francis Xavier identified 450 years ago that children essentially turn into their caregivers when he remarked with foresight that you shall "give me a child for the first seven years, and I'll give you the man."

Let's dig further into this concept of what happens during your first seven years of life. In just the past few years, scientists have identified that between the last trimester of pregnancy through the first seven years of life, a child's brain is in a sort of hypnotic state called theta. It's determined to be a low-vibration state below consciousness or awareness. According to Dr. Lipton, this theta brain wave state is a gift our biology gives us for survival. We are, hereby, programmed to download the rules, religion, regulations, language patterns, beliefs, indoctrination, and survival mechanisms of our families or caretakers. These neurological programs, rules, religion, indoctrination, regulations, language patterns, beliefs, and survival mechanisms help us function within our "tribe" and adhere to their rules. Remember: the brain is focused on survival above all else.

Dr. Lipton compares this childhood time to a computer program. During our first seven years of life, we have downloaded programs to our hard drive. All the while, we grow, mature, and reach adulthood with these programs running the show.

To rewire our minds, create new programs, become conscious co-creators with our Source, and to become essentially "reborn," we must interrupt the patterns that have gotten us this far.

To do this, you must learn to operate from the empowered, conscious mind. Consciousness is a term that is thrown around in an ordinal way; a person may claim to be "woke" or "conscious." When I hear that term, I immediately associate it not with a linear "better than you" status, but with this concept of mindful awareness of one's thoughts and actions.

The powerful manifester is consciously aware of their subconscious programs. Manifestation practices such as those in the Manifesting Secrets program are designed to interrupt subconscious programs every day to form new neural circuitry that we consciously choose to become, alas, subconscious, or automatic.

There are many things you've done today to interrupt your subconscious programs and set you on the path toward a new life through neuroplasticity. I mentioned earlier that merely by picking up this book, you've already initiated the neuroplasticity process. Exposing yourself to something new is, in fact, a foundational prerequisite to rewiring your thoughts and brain.

One French man is attributed with the greatest concentration of marketing genius and influence of the 20th century. Few know about him, although his marketing is among the most pervasive in the world. He contributed to the creative design for products in over 200 companies, including refrigerators, cigarette packs, cars, and even spacecrafts. His name is Raymond Loewy, and his theory was that to sell anybody anything, you must incorporate a blend of something being novel and yet not too unfamiliar, or "new and yet familiar." This principle is famously referred to as the MAYA principle: Most Advanced Yet Acceptable. He found that adults will embrace something new so long as it isn't too vast a departure from that which they already have accepted as normal.

Neuroplasticity is a game that can be played by these same rules. Learning something new compels you to "flex your mental muscles;" it's an exercise for your mind. When you read books about new concepts or speak aloud affirmations that take you outside of your normal thinking or box, you activate neuroplasticity. New things interrupt your subconscious patterns and prime your brain for new neural networking. When you do something new, you "shake things up," putting your brain on high alert.

Nevertheless, your brain resists things that are too unfamiliar. One of the best explanations I gleaned is from Dr. Mihaly Csikszentmihalyi. In his studies on the concept of flow state, he found that the brain achieves a heightened state of awareness and satisfaction when engaging in challenging tasks. However, if that task is too challenging, the brain becomes stressed.

The key to manifestation is to gradually make the unfamiliar become familiar to your mind. Meditation is a great example of a practice that deeply transforms the brain and primes our neural networks to create new synapses and circuitry. Just as we learned that a habit can improve 3700% over the course of a year with daily improvement of just 1%, as we introduce new concepts to our brains, we must remember the words of Raymond Loewy: Introduce new things, but with gentle ease. If something is too novel, our brains and bodies are inclined to reject it.

Repetition of manifestation practices with ease and flow will elicit the most powerful results in a gentle, permanent way.

CHAPTER FOURTEEN:
The Power of Relaxation

"During the time of stress, the "fight-or-flight" response is on and the self-repair mechanism is disabled. It is then when we say that the body goes down and the body is exposed to risk for disease. Meditation activates relaxation, when the sympathetic nervous system is turned off and the parasympathetic nervous system is turned on, and natural healing starts."
ANNIE WILSON

If you take nothing else away from this book, remember that nothing good comes from a constant state of fight, flight, freeze, or fawn. While this high-alert, trauma-response state is designed to keep us alive, it has gone haywire in modern times. During the caveman days, being on perpetually high alert served us because we lived among the wild animals with veritable threats around every corner. We didn't grab blueberries at the store, but we gathered them from bushes that may or may not have been poisonous. The threat of death was commonplace among our ancestors. Because of this, we have evolved to perceive innumerable things around us as threats. As Dr. Dawson Church explains in his book *Bliss Brain*, scientists have pointed out that our brains respond to to-do lists with as much fright as if there were a Saber-toothed tiger stalking us.

Training the brain to be at ease isn't just tough; it's an effort that goes against thousands of years of evolution. The brain, once again, is focused on keeping us alive; not thriving, much less happy. For the brain to change, we must be in a state of relaxation called the parasympathetic nervous state.

When our bodies are in a state of flight, fight, freeze, or fawn, we are unable to learn, grow, or properly heal. That's why when you're stressed, you can't seem to get that crick out of your neck or that aching to dissipate from your lower back. When you're on high alert; your body only functions from the sympathetic nervous state (SNS), during which non-vital functions shut down. In fact, when in the SNS, the blood rushes to your arms and legs, giving you the physiological upper hand to fight or flee danger. This, however, also means that blood rushes away from your internal organs, where your body focuses on resting, digestion, and healing.

Remember this simple description to help you remember the Parasympathetic Nervous State (PNS) vs. the Sympathetic Nervous State:

SNS: Fight or Flight
PNS: Rest and Digest

To become the most powerful in your manifestation efforts, you will need to become increasingly aware of how fight, fight, freeze, or fawn feels in your body.

Relaxation and ease are not passive states by any means. You may think that while you're kayaking in paradise or resting in the sun on your patio that your brain is "turned off." However, Functional magnetic resonance imaging (FMRI) machines reveal to us that it is during these states of relaxation that our brains are the most active and receptive to developing new programs and pruning programs we choose to eliminate from our brains. In a later chapter on Meditation, I'll describe this process with more detail, including tasks you can begin performing today to invoke this process of rest and digestion, neuroplasticity and, ultimately, epic manifestation.

CHAPTER FIFTEEN:
The Science of Happiness

"Happiness fuels manifestation."

I've berated the brain a fair share today when accusing it of not being invested or interested in our happiness. It's nothing personal. Your brain is wired for one thing: survival. But give your brain some credit. This magnificent organ perceives millions of bits of data per second. It doesn't always have time to stop and consider your happiness. Humans are evolutionarily wired for survival, not peace and happiness. Elevating your consciousness to pursue higher states of enlightenment such as joy, compassion, and non-judgment go against the grain of our brains. They require tremendous effort and conscious rewiring. In fact, not only are our brains wholly wired to promote survival, but we're actually evolutionarily wired to be discontent. When you're unhappy, you're more aware of threats to your survival. Brain scientist Dr. Dawson Church points out that humans are, alas, designed to be unhappy.

This isn't daunting. It's a challenge that you can approach with hope and optimism. Perhaps, like me, you've struggled with mind-numbing depression. In the next section, we'll learn about how to hack the brain for happiness.

Happiness is something you must train your brain to feel. There are many temporary types of happiness that are induced through chemicals (either natural or unnatural). People who have obtained unhappiness or depression programs from their caregivers tend to be particularly unhappy. People who were blessed with genuinely happy caregivers learned about how to be happy in a natural, more effortless way.

If happiness doesn't come naturally, it is still possible to manifest through dedicated training and rewiring. You can build happiness habits once you understand how to hack the brain for happiness. What's the science of happiness? When you feel good, your brain releases happiness chemicals such as dopamine, serotonin, oxytocin, and endorphins. These chemicals are designed to reward us when we do something that increases our survival.

First, let's discuss dopamine. Dopamine is our brain's reward chemical. It feels good and we're wired to seek more of it. All mammals produce dopamine. As we approach a reward, such as hunting for dinner, dopamine surges. The striving is the part that makes us happy; once we achieve a reward, the excitement of dopamine stops. That means that once you "have it all," you aren't necessarily happy. Your brain doesn't release those happiness chemicals anymore. In other words, you have to "do more" to "get more."

When you were a child, your brain obtained dopamine or pleasure pathways. In order to achieve dopamine surges today, our brains rely on those pathways and programs that provide our brains with a blueprint to dopamine release. We will follow these "maps" and seek happiness in ways that worked to bring us joy previously. Although we can rewire our minds to find happiness in new ways, it takes a lot of repetition.

That's right, there is neurologically almost no probability of rewiring the mind and manifesting a new reality without this critical factor: time. Repetition, building new habits, and executing on them daily is the way we create new paths in our brains. Remember, the brain was created for survival, not pleasure. Therefore, it is not "natural" to feel nonstop pleasure. As we age, we can consciously choose to rewire our minds to achieve happiness in newer and healthier ways. For instance, if you formerly found pleasure in binge eating or drinking, you can reprogram your mind to take pleasure in something healthier such as meditation, hiking, playing with your dog, or practicing martial arts. Is it possible to have an instant transformation? In fact, yes, although it's highly improbable.

If you experience a moment of euphoria or a sort of spiritual out-of-body experience, you can instantly rewire pathways. That's because highly emotional experiences engage neuroplasticity quicker. Rather than risking the mere possibility of rewiring your mind through inducing highly charged emotional experiences, the sure-fire way to rewire the brain to achieve happiness in healthy ways is through repetition. As discussed in the previous chapter, neuroplasticity is loosely defined as this: "neurons that fire together, wire together." Neurons, or nerve cells, receive sensory input or information from external stimuli. The first time you do something, you aren't usually very good at it because your neurons have weak connections; they aren't communicating together effectively yet; or "firing" together. With repetition, these neurons build stronger connections, i.e., they "wire" together. The art of manifestation is to do the right things so many times that these neurons fire together automatically or "subconsciously." Your subconscious mind controls approximately 95% of everything you think, say, and do.

When you assess a life that is unsatisfactory, it is because you have created that life with your subconscious patterns. In Part Three of this book, you will learn how to rewire what you want into your subconscious habits so that manifestation becomes automatic and more effortless. My Manifesting Secrets students have 90 entire days of various exercises designed to wire the mind for happiness.

Critically, happiness is a whole-body experience. One of the reasons Manifesting Secrets helps people change their brain programs quicker than other programs is because we incorporate the use of all three executive functions of the brain. These are your auditory, kinesthetic, and visual functions. Neurons fire more often when you engage all three executive functions of the brain. Therefore, they wire more quickly. Rather than using neurostimulation devices, Manifesting Secrets teaches us that everything we need to manifest for our dream lives can be done without the use of devices, gadgets, or visits to neuroscience specialists.

Meditation and Happiness

"I am not saying this because I am in need, for I have learned to be content whatever the circumstances. I know what it is to be in need, and I know what it is to have plenty. I have learned the secret of being content in any and every situation, whether well fed or hungry, whether living in plenty or in want. I can do all this through him who gives me strength".
PHILIPPIANS 4:11-13

It is said that your thoughts are 400% more influential on your happiness than external circumstances. You have heard that the Buddha spent hours meditating and starving himself silly, but you're probably also aware that he is known to have been supremely happy. So, too, can you look at Jesus Christ or the New Testament prophet Paul (among so many others) who said he could be content in any and every situation (Philippians 4:11-13)?

One thing that the world's happiest people have in common is undoubtedly this: a meditation practice. So why doesn't everybody meditate? Thankfully, the practice of meditation is on the rise worldwide.

"According to a 2017 U.S. survey, the percentage of adults who practiced some form of mantra-based meditation, mindfulness meditation, or spiritual meditation in the previous 12 months tripled between 2012 and 2017, from 4.1 percent to 14.2 percent. Among children aged 4 to 17 years, the percentage increased from 0.6 percent in 2012 to 5.4 percent in 2017.. Worldwide, it is believed that between 200 and 500 million people meditate. Some people find that number to be staggeringly high. As for me, the number is drastically lower than I would hope. Nevertheless, it's a remarkably sizable percentage of our global population."

-NATIONAL INSTITUTE OF COMPLEMENTARY AND INTEGRATIVE HEALTH-

One of the main reasons that meditation can be a struggle for people is because they don't understand how to do it. While I'll provide more examples later, it's important to know that when I refer to meditation, I am typically referring to seated meditation. However, meditation doesn't have to be sitting still for hours. It can include walking, reciting mantras, chanting, prayer beads such as rosaries or malas, focusing intently on one object, or simply watching the breath.

I suspect that one reason meditation hasn't become more widespread is because our reptilian brains are believed to be 430 million years old. These brain regions are those party-poopers who aren't interested in pleasure, but survival. Your prefrontal cortex, however, is only three million years old. It receives less energy than other brain regions that have evolutionarily pushed their way up the proverbial food chain of importance. Your prefrontal cortex covers part of your frontal lobe. It is designed to assist in complex cognitive behavior, personality, goal setting and achievement, decision-making, and regulating social behavior. Most importantly, the prefrontal cortex is tasked with helping you to focus your attention. Understanding this makes it clear as to why focusing our attention and meditating tends to get left until the end of the day... and often forgotten or put off until tomorrow. Evolutionarily, it plays second fiddle to survival. Conscious, dedicated habit-forming is the only way to wire our minds to seek meditation with ease, urgency, and effortlessness.

Remember what we learned in Chapter Seven about the two critical maxims of neuroplasticity: Number one: neurons that fire together wire together. When you create a daily consistent habit of meditation, that's when your brain becomes wired to make meditation a part of your life. The brain regions that get the most use are those which signal faster and get bigger. However, what you don't use, you lose. Meditation practices that aren't wired into the mind through daily, consistent habit will become pruned with other rarely used programs.

How you meditate is measurably important for its effectiveness. Scientists have found that meditating for hours on end isn't always more effective than meditating consciously for five minutes. People trained in meditation can harness alpha and gamma waves in minutes versus hours.

Why is meditation important for happiness? Meditation changes how the brain works. During meditation, certain brain regions work harder, and others do not fire as much. The assumption that everything slows down during meditation is entirely wrong. As you'll learn in Part Three, the brain actually lights up with activity when we meditate, or any time we are in a state of deep rest and relaxation. This is why you suddenly think of something important in the shower or come up with your greatest poetry while lying in bed, nearly asleep.

When you meditate, your amygdala, which processes fight or flight, and parts of your parietal lobe, which processes sensory input, shut down. Don't worry, it isn't like your car is running out of gas; it's perfectly safe to enter into this state. Your body is in its most powerful capacity to learn, grow, and heal when it doesn't perceive threats. Quieting the amygdala and parietal lobes are critical for states of deep learning, growing, and healing.

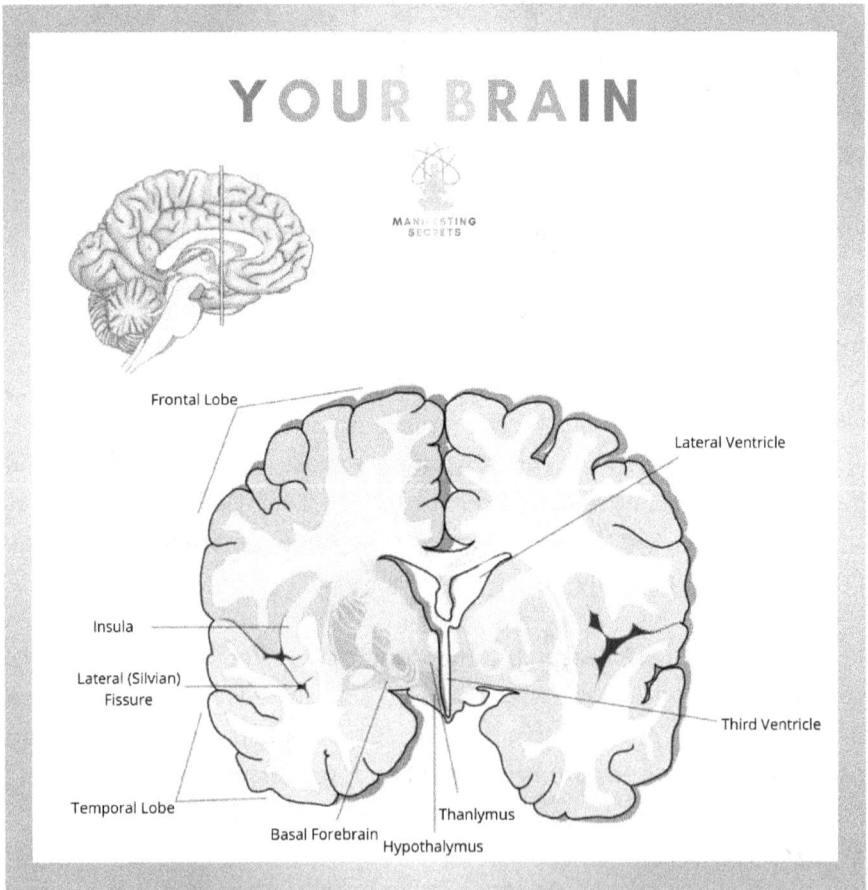

YOUR BRAIN

MANIFESTING
SECRETS

Frontal Lobe

Lateral Ventricle

Insula

Lateral (Silvian) Fissure

Third Ventricle

Temporal Lobe

Thanlymus

Basal Forebrain

Hypothalymus

Other parts of the brain light up during meditation, such as the prefrontal cortex, that region of the brain that helps you focus your attention. Most interestingly, the regions of your brain that synthesize your personality go dark during meditation. The inner critic shuts up, and you immerse with a gentler, loving, expansive universe. Meditation helps you get out of your own way by driving the self-critical "you" out of your mind. You can see why it's so important for manifestation. Once you turn off your inner critic, you more easily tether yourself to feelings of oneness, connectedness, compassion, and love. And if you've learned anything so far about manifestation, it's a practice of love, patience, compassion, and kindness.

We have learned through FMRI machines that the parietal lobe goes dark during meditation in experienced meditators, such as monks who have thousands of hours of meditation behind them. Empathy, relaxation, compassion, forgiveness, and self-esteem are all attributes of a quiet parietal lobe. Dr. Dawson Church illustrates in his book *Bliss Brain* that when these subjects' brains were studied, it was found that parietal lobe activity dropped by up to 40% as they entered an altered state of consciousness during meditation. To put it in perspective, on average, the fluctuation is only about 5%.

NEUROCHEMICAL EFFECTS OF MEDITATION

NEWBERG AB, IVERSEN J
MED HYPOTHESES. 2003 AUG: 61 (2)282-91.

Neurochemical	Observed Change
Dopamine	Increased
Serotonin.	Increased
Melatonin	Increased
DMT	Increased
Noradrenaline	Decreased
Acetylcholine	Increased
Glutamate.	Increased
NAAG	Increased
GABA	Increased
Cortisol and CRH	Decreased
AVT	Increased
β-endorphin	Increased

MANIFESTING
SECRETS

Another fascinating way that meditation functions is to help us connect with others. Although meditation is an independent endeavor, it deeply strengthens our bond to the collective, or to humanity as a whole. We are social beings designed to connect. One hundred forty-eight studies in a meta-analysis with a total of 308,849 participants revealed that those with strong social relationships enjoyed a 50% stronger increase in survival. This finding was consistent among various age groups and demographics. In fact, avoiding smoking, obesity, or a lack of exercise have been found to be less important to life expectancy connectedness to others.

Connection with others and meditation share one critical commonality. They usher us into the present moment. In the present moment, we are less plagued with guilt or worries about the future; we transcend future worries.

Using advanced FMRI machines, researchers such as American neuroscientist Andrew Newburgh, who studied Franciscan monks and Tibetan nuns, says that "when people lose their sense of self-feeling a sense of oneness, this results in a blurring of the boundary between self and others with no sense of space or passage of time." When you meditate, you tend to lose your sense of self in favor of a sense of universal oneness. As you meditate more, you may begin to feel like you've taken the red pill and you're coming out of the Matrix. Experiences with clairvoyance, precognition, and other encounters with a Divine Source become more common. Eventually, the real world doesn't seem quite as real as the unity consciousness gained through meditation and the quieting of your parietal lobe.

CHAPTER SIXTEEN:
Smiling and Manifestation

"A smile is the most attractive outfit you can wear."

How does smiling manifest prosperity? It's a butterfly effect, and I'll detail the process with references to science to prove it. Smiling changes your mood in an instant. But what if you're in a bad mood? Is smiling dishonest? Is this trite advice from a teacher who's popped way too many positive thinking pills? Nay. Great teachers, sales (wo)men, and epically successful billionaires swear by the science of smiling. Imagine a photo of Richard Branson or Tony Robbins in your head. Talk about infectious smiles.

Smiling improves your mood instantly and can turn the sour into sweet. I remember once shortly after my son was born when I was driving to Denver, Colorado listening to Dale Carnegie's "How to Win Friends and Influence People." Something Carnegie said stuck out and called to me. He said that smiling is like a ray of sunshine, simple as that. He discussed how our happiness is contagious, and that even if we don't feel like smiling, we can become happier by doing so.

At first glance, it sounded like Carnegie was encouraging people to be disingenuous. Still, I became more liberal with my smiles from then on. I hoped to share joy with people around me... and I really wanted my son to feel the comfort children have when his mommy was happy. I then began incorporating smiling into my meditations. When I would get stressed, anxious, frustrated by temper tantrums, or even when I felt self-conscious or self-critical, I'd close my eyes, smile, and tell myself I'm safe and beautiful. Like clockwork, my mood would improve. My smile helped serve as a neurological and psychological trigger to remind me that everything always works out for me.

A few years ago, I attended a wedding and some of the guests chastised me and my business. I felt humiliated. When I came home, my stomach was bloated (likely due to Sacral Chakra trauma or past traumas being unearthed), my face broke out in horrific acne, and I felt depressed, tired, irritable, and distraught about everything. I don't let the blues get to me and have always been fantastic and bouncing out of dark times, but this darkness seemed to last for weeks. I couldn't get back on track; I was traveling for work every week and feeling very codependent, giving so much more to others than to myself.

The pinnacle of this dark time was when I attended a meditation and yoga retreat in Moab, Utah. My acne was so bad that I couldn't smile without tearing the skin on my cheek. So, I didn't smile. I grinned a little, but I was in pain inside and out. After many hours in yoga and meditation, my mind and spirit started to transform. I felt myself smiling more on the last day of the retreat than I had in weeks. It occurred to me that I hadn't been smiling much because of the acne or because of my traumatic family experience. Suddenly, a switch went off and I saw how inseparable my smile is from my happiness. It's not only the outward sign of my inner peace, but it's a muscular trigger for increased happiness and freedom from anxiety.

I allowed the traumatic wedding event to kill my joy. It literally manifested into cystic acne that prevented me from smiling. And suddenly, when I forced myself to smile, I began to feel my spirit's lift. That afternoon I parked myself on a rock at the edge of a river in a huge canyon in Moab near where one of my favorite Corona Arch hikes begins. As I laid in the sun and meditated, I forced myself to smile throughout the entire meditation. I said the words, "I am light. I have a purpose. I have meaning. My life is a smile. I bring joy to others. I am the light of Aspen." I baked in the sun and meditated on my heart's mission in life; not thinking about my failures but focusing on my being. As I glanced down that river, I saw myself in the river and suddenly had a vision of my work being like that river; feeding so many tributaries; which were the businesses I would serve in my coaching and book publishing programs. Until I forced myself to smile and say these mantras with faith, I didn't feel the joy that I needed to manifest new abundance, a second wind in my business, and even the healing energy I needed to beat the cystic acne once and for all. In the next few days, the acne (which I had for nearly two months) disappeared. By relaxing and smiling, I allowed my body the state of relaxation it needed to turn back on my own self-healing superpowers.

After that week, I began digging into research about how smiling lessens pain, triggers happy hormonal responses in your brain, and even causes feelings of happiness. I found that frowning causes anxiety in your brain, but that when you smile, you use muscles in your face that trigger the response of dopamine in your brain, one of the happiness chemicals we discussed in the Science of Happiness Chapter.

Dopamine is a neurotransmitter for your brain's reward and pleasure centers; the more dopamine something triggers, the more happiness your body feels. Dopamine allows you to feel bliss and euphoria. Being unable to physically smile for a few weeks, I learned about how important smiling is to my own personal happiness. As my son grew up and is nearing 6 years old at the time of this book, I've used this method of smiling when faced with anxiety. More days than not, people ask me how I can be so patient with my son, smiling in the midst of tantrums or trials. I tell them frankly that my smile isn't what I'm feeling sometimes, but it gets me to the feelings I want. I believe that smiling is a gift from God; He even rewards us for doing so.

My smile is sunshine. This has been one of my regular mantras for nearly four years now.

CHAPTER SEVENTEEN:
Manifesting Money

"Abundance is not something we acquire.
It is something we tap into."
Dr. Wayne Dyer

Manifesting money is a topic for another book entirely, but because this book is the first in the Manifesting series, we'll provide a cursory overview of ways I have found to be effective at manifesting money and prosperity. Naturally, as I increase my wealth and reverse engineer the steps that were most effective, I'll continue to share stories.

What I can guarantee you is that whether my wealth exceeds $100,000,000 or $1 billion, these principles are hailed universally by wealth manifestation masters.

This year I've already quadrupled my income from January of 2020 through many of these principles; and they're so simple that you can begin incorporating them today.

Ask

Most fundamentally, ask not and you will receive not. A few months ago, I called my son's school and indicated to them that having the children working from home (for only an hour a day) was hindering my ability to work. I requested that childcare arrangements be made through the school so that single parents like me could continue to take calls and keep my business running efficiently during the endless quarantines.

Several months later a representative from a family care group in Aspen called me and indicated that an anonymous donor wanted to gift me $6,400 that I could use toward childcare or other arrangements with my son out of school. I was in shock. I called hoping to find playgroups, after-school programs or cohorts, but the director chose to bless me with this incredible gift.

When I moved into my apartment in Aspen, it was dilapidated, run down, and replete with countless broken appliances and fixtures. For a $1.75 million apartment, this was unacceptable. I asked my landlord if she would reduce my rent by $1,600 monthly so that I could invest in renovating the place to acceptable standards. As of today, I've manifested well over $10,000 from that simple "ask."

When I launched my Manifesting Secrets program, I was bold in asking friends and peers to promote my new Facebook group and Facebook Live events. I was floored to see that in just one week the attendance for my live event and participation in the private Facebook group was more than double anything I'd seen after three whole years in my previous company, working day and night. Ask and you will receive.

Along the same vein as asking others for what you want is to ask the universe for what you want. I regularly pray for my son, my family, my prosperity, and my inner righteousness and consciousness at night. When friends stay at my house, I will spend time giving them massages and praying for them or supplicating on their behalf with my higher power, whom I call God.

If you're a parent or a lover, you understand how hard it is to say no to somebody you love. The universe in its infinite abundance does, indeed, host every desire you could ever conceive and every provision to materialize that desire. Just like you rush to fulfill the wishes of your child or lover, does not the universe rush to deliver your desires? How do you know? Have you asked?

Happiness

In a previous chapter we discussed the science of happiness. How does it transfer to the acquisition of wealth?

As you may have seen, happier people manifest wealth more easily. When I was depressed and downtrodden, I used to look at my optimistic younger sister and think, *How could that sweet, sappy, Pollyanna little brat be accumulating so much coin?* As the more cerebral, existential, and emotionally unstable sister, I was infuriated.

As I studied deeply the art of happiness and meditated for a few years, I learned how happiness attracts wealth. Indeed, I believe that what we feel we attract and that our positive vibrations beget positive manifestations. But even deeper, when we are happy, we manifest from a place of more creativity, love and excitement. When we are happy, we are in a state of present moment awareness.

If you're selling anything, which (newsflash!) we're always selling, people want to work with a happy provider. Your positive energy infects others. In fact, a smile can increase conversions on the close you seek to make either with a deal or a sale that will bless your bottom-line abundance.

Present Moment Awareness

Present moment awareness is like a fast-track ticket to quantum manifestation. You may think that worrying about your future will help you manifest more money. Rather, being a mindful steward of the present moment will keep you aware of the blessings right in front of you. Somebody recently asked me what the opposite of fear is? I don't know if it's the opposite of fear, but it's certainly the anecdote: and that's present moment awareness. Present moment awareness is where the past traumas we've experienced don't hurt us anymore and the future worries don't paralyze us. Rich people don't fear. They operate with faith and confidence in their ability to get rich.

When you run your business or financial life with present moment awareness, you become more courageous. You make decisions less on what negative things may happen and more decisions based on what you want, which are in alignment with your highest self and true life's calling or purpose. When you invest in your business or life with present moment awareness, you listen to the signs, synchronicities and nudges from the universe that lead you to your highest path.

Okay, I know a lot of this sounds pretty wu-wu. So, let me give you an example. One of my client's was sitting in front of her computer the other night reading some emails. One email came in from a company she follows: a publicly traded media group. The email mentioned that there were job openings in sales and marketing for this company. My client, having run sales teams for over twenty years and having sold multiple tens of millions, perhaps even hundreds of millions in products, emailed the company. It sounded something like this:

"Dear sir or madam,

I've been following your newsletter and enjoying the content you produce for two years now as a member of your company and am an avid follower. I would be remiss if I didn't add my two-sense as an experienced sales coach having run multiple sales teams for some of the world's most esteemed investors. I lovingly suggest you try ___, _____, and _____ to attract and hire your ideal sales manager. Once you find him or her, I suggest you _____ and _____. Yours truly, Sarah."

Within hours of her email, she received a response from one of the company representatives. "Sarah, may we call you?" Sarah, wanting to give from her heart and love for the company and the conscious productions they had curated for her pleasure, answered affirmatively. "Yes, I'll be honored to help you out."

Within weeks, Sarah had an offer from the company that was so appealing that it allowed her to sell or even fold three of her companies just to join this one. The CEO moved her office right next to him.

I asked Sarah after we squealed and giggled for about 90 minutes, "Sarah, what did you do to manifest this absolute dream job?"

"I was fully present in the moment." She answered. "That month I meditated each day for 30 minutes at night and in the morning in addition to an hour at lunch. I've been present with my body, eating vegan for the past year and listening to my body's sensations. When I got that email, I absolutely had to respond; I didn't think about the tea kettle on the stove or a lot of emails I hadn't yet answered. I was there, with that email, with my whole heart. And I was present with my future self. I was in alignment with her. I wanted to move to the mountains by age 41. I kept visualizing that perfect world, but I was too caught up in my life on the East Coast to move West... That is, until I got an offer I couldn't refuse. Do you know something that you told me once that I never forgot? I never arrive. I'm always exploring. When I went into that meeting, they asked me questions for four days about who I am and what I do. I committed to this: I am an explorer. I'm not living in the future or living with past domestication or indoctrination. My destiny is my search for truth in the here and now."

Leverage Your Assets

To manifest epic wealth, it's critical to know your worth. You have a superpower. You have a calling and a gift that people will pay for. I ask my clients to draw a ladder in their mind's eye. At the top of the latter is your hero; maybe it's Deepak Chopra or even Jesus Christ. In between you and that hero, there may be two, three, or even twenty levels of wealth and abundance, and the remarkably different downloads and wisdom you need to ascend to the next step. Who are you? You're perfect, just where you are. You are ascending to greater levels of consciousness and skills to obtain to get where you want to go. However, there are millions of people just two steps below you who'd love to get where you are today.

When you understand your worth, you will find that you see clearly the very hands you're destined to hold to lift up to your level. Those are the very target clients who will invest money to get where you are. If you're a hairdresser, attorney, engineer, or yoga instructor, these principles apply. In our society, it's nearly impossible to become rich by saving yourself.

Most people live paycheck to paycheck. However, if you design a side hustle around this concept of getting paid to teach others what you know, you'll find that you can often increase your income tremendously in just a few days or weeks' time. You can keep your nine-to-five and still earn more money by knowing your worth. In fact, did you know that in a study of over 160,000 participants, 70% of those who asked for raises were granted one? Sometimes we don't receive because we don't trust our worth and then, as we stated in the first money manifestation tip, just ask.

In the stories you'll read in Part Three, you'll see how I asked for promotions, money, and discounts to manifest tens of thousands more dollars into my life almost overnight.

Set Money Manifestation Intentions

Here are the money manifestation intentions that I keep on my refrigerator door. Head over to www.manifestingsecrets.com/bonuses to download your free Money Manifestation audio today. You can even listen on your phone while you're making breakfast or driving to work. Here are a few of my personal money manifestation mantras:

* Wealth is my birthright
* The Universe is conspiring to make me wealthy
* I deserve epic wealth
* I give myself permission to be rich.
* Creating wealth is fun
* I am joyful, creative, and expansive
* My superpower is worth billions
* I gratefully receive the magnificent flow of financial abundance that is coming into my life right now. (Four months after I began saying this mantra, my business quadrupled its revenue from the same month the previous year.)

Make Money Manifestation Spiritual

Manifesting wealth is your birthright; it's what you're designed to do. In line with the Law of Attraction, when you believe that creating wealth is hard, you manifest that. When you believe that creating wealth is joyful, expansive, and effortless, it will become so. If you believe it's hard work to get rich, the Universe will deliver that to you.

A great teacher of wealth, David Cameron Gikandi, details in his book *A Happy Pocket Full of Money* that when you create ripples of wealth around you, those ripples will come back to you. Invest in people you believe in. Tip your server well. Pick up the dime you see lying in the parking lot and say, "Thank you God for this blessing. I gratefully receive the magnificent flow of financial abundance that is coming my way, right now."

All your work is spiritual, and that includes making money. Your nine-to-five brick laying or hairdressing or doing construction is spiritual work. You are a spiritual being, so treat your work as a spiritual endeavor and you will receive quantum rewards.

Part 2

manifestation stories

CHAPTER EIGHTEEN:
How Manifestation Looks In "Real Life"

"God answered, I will be with you."
Exodus 3:12

Now that we've immersed ourselves in the scientific fundamentals of manifestation, I want to have some fun in Part Two of this book before we head into the seven critical steps you can take to start manifesting today. I intend to publish a new book every year in The Manifesting Series; with greater, more shocking, more abundant, more self-aware, and more magnificent stories of manifestation as my own manifestation skills improve.

The primary purpose in sharing these stories is to illustrate to you how delightful manifestation can be; and how simple it is, as well. Frequently, people share with me that they see manifestation as crystal balls, tarot cards, and mystical experiences. Excitingly, manifestation isn't generally very theatrical. It's the result of daily, habitual manifestation exercises that essentially grow your manifestation muscles.

Remember the two principles of neuroplasticity: "neurons that fire together, wire together" and "if you don't use it, you lose it." In these stories, be on the lookout for how I leveraged a powerful daily manifestation practice with self-awareness, confidence, and faith to generate income, happiness, and healing. As I "fired and wired" neurons that benefited my manifestation goals, I strategically "lost" negative thoughts and behaviors along the way.

Manifesting has enabled me to live in a deeper expression of myself with more satisfaction and happiness. Largely, as we read earlier, happiness from manifestation hasn't come from material things, but from contentment in what I already have and have already manifested. You won't manifest something you don't first visualize.

You'll see in this collection of stories that contentment with who I am and what I have is just as pleasing as prosperity. It's important to me to illustrate that my manifestations aren't private jets and yachts, yet. For some of us, the focus will be on a billion-dollar empire. That goal is appropriate for some, and not for others. Quite simply, there are many of us who will never manifest that amount of wealth in this lifetime. There is far too much focus in the manifestation and personal development community on manifesting money.

While I believe that manifesting financial security is laudable, it isn't illustrative of one's manifestation progress by any means. And I will be the first to admit that not everybody is destined to be a billionaire. Far too many eager manifesters give up on manifestation when they don't manifest epic wealth. Focusing on the dollars is, alas, missing the point. Inner peace is the desire of countless billionaires and largely the focus of my personal manifestation efforts. I'm happy to say that prosperity indeed increases with my focus on inner peace. As my inner peace decreases, my financial woes magnify. Every time.

For others reading this book, we'll be happy to meet our soulmates and live a simple, comfortable life with loads of free time that are spent on a river or in a mountain versus a stock portfolio. Those of us who manifest private jets are no better than those of us who manifest happiness. In fact, the latter is far and away a more desirable manifestation that more elegantly elicits financial prosperity. As much as some of us desire material things, it has been proven time and again that the greatest manifestation is love and contentment. Most often, material prosperity increases directly in proportion to internal happiness and satisfaction. I would argue that no Ferrari can bring greater joy than connection with others, non judgment, and divine encounters with the Source or God.

I still have bad days, and bad seasons just as any human does, although I know that I can shift and turn my thoughts quicker with each passing week to happiness. What's more, my financial prosperity becomes increasingly more reliable and magnificent. Above all, the stories you will read will illustrate that love is the greatest manifestation you could ever materialize. There are no limitations in your own manifestation power if you use the tools and resources from this book. I pray that these stories will give you hope and faith in your own manifestation success stories, which I urge you to share with the Manifesting Secrets private Facebook group at www.facebook.com/groups/manifestingsecrets/

CHAPTER NINETEEN:
Mountain Lions and Letting Go

"A truly strong person does not need the approval of others any more than a lion needs the approval of sheep."
VERNON HOWARD

I hit a low point in life when my son was still a toddler; every day felt, as Cheryl Strayed so beautifully expressed, like I blew through my entire love reserve. I had tried everything to manifest a better life, but I was trying too hard. I suffered years of turmoil with romantic and platonic relationships, single parenting was kicking my butt every day.

I had moved to a remote mountain town in Colorado and lived on a hill with my son in a neighborhood so rudimentary and off-grid that it didn't even have sidewalks for my son to play on. The home we lived in didn't have a yard; what yard it had was as steep as a black diamond ski run and covered with thousands of tiny cacti; putting one hand down in the dirt could elicit the sting of a hundred cactus prickles.

To add insult to injury, my businesses failed during those toddler years with my son, although I had hired staff members; invested tens of thousands into marketing and advertising with little-to-no ROI; I worked out like a fiend; I tried making new friends constantly everywhere I went; and I even joined high-level masterminds with people who were invested in helping me grow and prosper.

All the hard work resulted in nothing but exhaustion and a disharmonious relationship with my masculine and feminine natures. One day a man told me that I was intimidating; that I had so much masculine energy that I turned him off and made him feel threatened; like I was a live wire.

And if you were on the receiving end of some of my emotional outbursts, you had every right to be scared. I felt humiliated that I'd lost my faith, elegance, grace, and hope.

Although I have perspective now, at the time I couldn't understand what I was doing wrong. I was trying terribly hard to succeed; indeed, nobody on the planet could outwork me. However, hard work in excess isn't healthy, no matter what the "hustle porn" tries to tell you. I want you to understand this: when you work without play, fun, and even time frivolity, you actually destroy your manifestation power. What we feel, we attract. When you are always in a state of scrambling to work hard; the universe will respond in accordance. When you believe that you "must" work constantly and blindingly hard, you will manifest this life. It's one of the biggest reasons people stay poor; they work too hard and don't manifest from a place of joy, love, peace, and relaxation. Your brain is stressed by such hard work. When you exhaust yourself, you are rarely rewarded with great wealth, but more often cursed with burnout. During burnout you make less money and destroy the relationships and allies you need to manifest the life you want.

I didn't truly grasp this at the time. The harder I tried to dig myself out of a financial and relationship hole, the lonelier and more desperate I became.

It was during this time that I slipped a disk in my back and broke my foot, the stories I told in Chapter One of this book. I began to lean into meditation, self-healing, and other manifestation practices that have absolutely transformed my pain into pleasure and power. Seven of these practices are detailed in Part Three of this book.

In this story, we'll look at a meditation experience that served as a catalyst to great transformation in my life. One night I was meditating at the end of my bed with my best friend. I rarely open my eyes during meditation, however, at this moment I felt a gentle breeze from the corner of my room near the door. I opened my eyes and saw a vision of a mountain lion walking across my room, clear as day. As she stood at the end of my bed, I felt like I could touch her coarse fur with my hand. She was in so many ways "real;" as her very real energy stood there and materialized into this vision.

I had a lot of profound spiritual experiences in that house on the hill, ranging from visions to almost heart-stopping Kundalini energy while practicing yoga there to sensing and exercising a demonic presence with a friend. Indeed, the house was absolutely riddled with spiritual energies; some good and some bad. Perhaps it had nothing to do with the house; but with my willingness to see the spiritual world therein. My spiritual eyes were open even when my heart felt so battered and closed.

On this special night that I saw a mountain lion, I didn't know how to describe the phenomena of seeing a vision, spirit, entity or demon. I don't necessarily think that these types of experiences are meant for everybody. But since I was a child, I've always had a sort of sixth sense. I see things energetically and intuitively that sometimes materialize into visions. Sometimes these visions are merely a lingering presence that is ominous and seems to resist leaving. For instance, in a later story I'll share with you that I once "felt" that a man was a threat to children in a hotel, and later found out he was a child predator. Another time, I "felt" that a man was cheating on me, and without knowing anything about the situation, I looked at my lover and told him that I knew he was cheating on me, and I even had the woman's name.

The mountain lion who appeared to me appeared in a sort of slow flash. She seemed to saunter across my room slowly from the door to just in front of me where I'd been meditating on my bed. I didn't feel scared, but in awe. Almost as quickly as she appeared, this mountain lion vanished. I wasn't aghast or surprised. I just accepted the lion and her presence. She was slightly ominous and certainly large, but the vision came with a spirit of kindness, compassion, and respect. I knew immediately that she came to teach me something.

I turned toward Bel who was meditating beside me and said, "Bel, I just saw a mountain lion and she has a message for me, but I'm not sure what."

Bel reported that two mountain lions had just been shot above my home for breaking into a stable and eating a neighbor's llamas due to human negligence. We discussed the sad story for a moment, and I asked him if it was normal for mountain lions to prey on llamas. "No, not at all. Normally they have plenty of food either from roadkill or other animals in the fields up there," he said pointing to the hill behind my home. "But after the fires that decimated the mountains in the 2017 fires, the food is scarce. That's why the lions have moved down here near us; they're starving, and their homes have been destroyed. They're looking for food."

At that moment my chest felt warmth and I heard a voice in my head say, "Stephanie, I love you, but I need you to let go." Let go of what? I wondered. I looked out into the night from my bedroom window.

That's when I realized that I needed to move myself and my son from that environment. Like the mountain lion, I was starving and taking desperate measures to feed myself and my son. I was going to get hurt if I didn't migrate to a new home with more abundant sources of business, joy, and relationships. I had experienced a tremendous trauma in that house, and it was time for me to migrate away from that trauma both spiritually and physically.

And so, I moved to Aspen a few months later. Sure enough, the week I moved to Aspen was one of the most magical weeks of my life. I was honored with the title of Top 20 Personal Development Leaders to Watch by Business Insider and was featured in Forbes Magazine as an Entrepreneur Expert. I manifested tens of thousands of dollars in gifts and help from friends. I began to attract people around me who wanted to be friends. In my previous location, I was isolated and felt lonely and disconnected. Suddenly, the doors of prosperity opened, and blessings started pouring out from every direction. Within a few months, my business revenues quadrupled, and I began looking for three new full-time employees to handle the workload.

I used so many tools from my Manifesting Secrets to call in this miraculous surrender and abundance, and I'll describe more of them in the remaining pages of this book. Among so many, here are some of the most important ones.

1. I *Surrendered.* There's a science to surrender and it's detailed in Manifesting Secrets. I took away my own control and just voiced

my wants and needs, not knowing the how, why, or when. Read more about this in Part Three.

2. I used *meditation*. Meditation would calm my mind when the weight of perceived failure and imposter syndrome was suffocating me, tempting me to stop and just "get a job." You'll also learn some shocking ways you can leverage meditation to manifest a better life in Part Three.

3. *Relationships*. I used various relationship tools to manifest the people I needed to help me get to where I wanted; both from a public relations perspective and a social-emotional perspective. I think of relationships as our manifestation allies.

4. *Mentorship*. I used the mentorship of expert coaches and healers from the Manifesting Secrets course.

5. *Movement*. I moved my body with Brain Flow Yoga and other movement practices to clear pain and align my chakras so that I would feel grounded, wise, and powerful.

6. I used *affirmations* every day that rewired my mind for prosperity.

In my new Aspen life, I've enjoyed much greater prosperity and elevation in my relationships. However, I also choose to remember that special mountain lioness and her message. I keep several pillows around my house with an image of the lioness today.

CHAPTER TWENTY:
Manifesting Safety from a Child Predator

"Finally, be strong in the Lord and in his mighty power. Put on the full armor of God, so that you can take your stand against the devil's schemes. For our struggle is not against flesh and blood, but against the rulers, against the authorities, against the powers of this dark world and against the spiritual forces of evil in the heavenly realms. Therefore, put on the full armor of God, so that when the day of evil comes, you may be able to stand your ground, and after you have done everything, to stand. Stand firm then, with the belt of truth buckled around your waist, with the breastplate of righteousness in place, and with your feet fitted with the readiness that comes from the gospel of peace. In addition to all this, take up the shield of faith, with which you can extinguish all the flaming arrows of the evil one. Take the helmet of salvation and the sword of the Spirit, which is the word of God."
EPHESIANS 6:10-17

In September of 2020, an innocuous comment I made in the Aspen Times elicited a storm of legal backlash. At times I felt helpless, even gagged. Being quiet with such strong convictions as I have caused myself to feel like the world was burning and I couldn't protect anybody, much less even cry, "look out."

In 2020, I grew my business tremendously and labored to manifest the best possible year for Hunter with all of my Manifesting Secrets tools. Although the world turned upside down in 2020, I sheltered my sweet five-year-old boy from the terror that so many around us felt. At times, I even felt that terrors; staying awake so many nights dumbfounded, trembling, and wondering how many more punches we would take to the gut as a collective.

Still, I kept getting up; I kept pushing ahead; I kept innovating. I kept sowing love and I learned, above all, that the most loving thing to do sometimes has some serious warrior, mama bear vibes.

Spring and summer days were as sweet as you could imagine. We hiked for hours, visited caves, painted rocks, adopted cacti, swam in rivers, climbed and camped in the desert, soaked in the hot springs, and camped more nights than we stayed at home some weeks. We read for hundreds of hours. We meditated together. I meditated while he slept next to me in the twin bed and I slept next to him on a trundle bed in his room. I meditated so much that year. So much more than I ever imagined I'd want or need to. And yet, I want so much more. I am so grateful to have expanded my meditation, visualization, and breathwork practices this year. I will never be the same.

As the weather turned cold, we struggled to find things to do. The library wasn't open, and the ones that were had yellow police tape around the kids' sections. We played at the outdoor playgrounds until it became downright treacherous. We sledded almost daily until our legs were so sore it was hard to do anything much more than lie down at the bottom of the hill and make snow angels. I'm so grateful for the natural beauty around me and the gift of health to explore these surroundings with my boy.

One cold night toward the end of 2020, our natural gas company experienced a vandalism in Aspen that left 3,500 people without hot water or gas. We stayed home for the first 20 hours or so, but then gave up after sledding the previous night.

We headed down the valley to a hotel with an indoor pool. If our house is freezing, I wanted to make an adventure out of it. I surprised Hunter with a last minute "vacation" to Glenwood Springs. He was beyond elated.

We arrived fairly late one night, just in time to get in an hour of swimming at the pool. Hunter must have told me 200 times that it was the best vacation ever.

One afternoon we swam and played all day at the pool until Hunter was so tired, he needed to return to our hotel room and crush an entire mini fridge worth of snacks I'd brought him. We planned to return to the pool that night, but my stomach felt sick about it. Suddenly, I felt like I was in the wrong place at the wrong time, like we didn't belong in that hotel.

As Hunter and I enjoyed a few hours of stillness and rejuvenation before our evening swim, the sick feeling I had encountered began to swell in my belly, almost to the point of making me a bit nauseous. Have you ever had a feeling that your lover is cheating on you or that somebody you love has been in a car accident? "Tonight, something is happening," I told myself soberly.

I prepared with meditation and movement before we went down to the pool. I held light in my hands and scrubbed it all over my body. "I am light. I am light." I chanted. Because my body prepared me for something (I wasn't sure what, yet), I visualized various scenarios and how I could hold my cool and use emotional regulation. I meditated with my spine straight and visualized light raining down around me. Still in a meditative state, I pushed water around me as we soaked in the hot tub, imagining light emanating from my body, infusing the whole room.

Sure enough, a disheveled, menacing dude found his way back into the pool that evening. I immediately understood that my intuition had prepared me to protect the children in that hotel; I knew that the man was a child molester.

Within 30 minutes of his arrival, I observed that there was a camera under his sweatshirt, and he was talking to children who weren't his. In fact, I asked him if he had kids. When he said that he didn't have any, in the middle of an indoor kids' water park, I was ready to call the police. I waited until I low-key invited six or so other parents to share any weird behaviors they saw and agreed to share them with the police.

Within the next few moments, I learned that people saw him taking photos of me and my son, drinking booze from a backpack that he had, bragging to young kids that he had a motorcycle that he wanted to give to a special little boy, and then, the straw that broke the camel's back, he told me he wanted to give my son candy. I gave one of the mothers I had enlisted in my detective work the "eye," and she went to the front desk to ask them to call the police.

Hunter did a run down the water slide, with me catching him at the bottom, and went to run up the ladder to take another run. No sooner did I turn my back to ready myself at the base of the slide did that creep yell for Hunter. "Hey little guy, I have some candy for you." Was this really happening?

I turned just in time to see him reach out his arm and immediately yelled, "Hunter, don't you touch that." Hunter was mortified. I was righteously and insanely pissed off. I asked Hunter to run to the slide again while I pressed record on my phone's video and confronted him.

The cops arrived, and within a few minutes I learned that the creep had caused some trouble in another Glenwood Springs venue just a few nights earlier. He had gotten away with hanging around this new venue somehow for a few days. Parents who'd arrived Sunday corroborated stories from others who, like me, just arrived.

I looked around and wondered: how could a dozen families all stand here and watch this pervert from different angles and not do anything? It appeared to be a living analogy of the country that slid into such disarray and economic turmoil during the previous year.

Some of the parents I approached had shared with me that they were uncomfortable that he was flirting with the pre-teen girls. Others were uncomfortable that he asked them to enter a locked pool area. Others saw him carrying booze. Another dad saw him taking photos of me. One mom said the man was knocking on her door when she went to grab dinner, telling her two young girls that they'd receive candy and presents if they opened.

But everybody looked away.

That is, until a powerful manifester came in.

I believe that Manifesting Secrets is one of the most powerful manifestation programs on the planet precisely because our students learn to harness the power of their bodies. With that, you will heal and align your energy centers known as chakras. You will learn to put your body into a state of self-healing. You will also deeply enhance your intuitive awareness.

As a powerful manifester, your body or "gut" will tell you when to back down and when to fight back. You will learn to be okay with feeling scared, suspicious, and angry within reason. While you won't let these emotions take control of your life, you will learn through Manifesting Secrets to harness those emotions and transmute them into power and self-protection. And if you're lucky, you may even be able to use your emotions to protect others, like I did on this remarkable day.

CHAPTER TWENTY-ONE:
Manifesting A Child

"I manifest things I want by believing them into existence."

I had a tough time conceiving a child. I blamed this in part due to the eating disorders that plagued my 20s. I also believe that my difficulty in conceiving also derives from the pain I was in from previous abusive relationships as well as my subconscious fear of bringing a child into the world with ill-chosen partners. The trauma in my body manifested into three miscarriages between 2010 and 2012; one was second term and I had to have surgery to remove the baby from my womb. I was throttled by these tragedies and mourned for many months with depression and the feeling that I was a bad person because my body kept killing my babies.

What helped me heal was to recognize that God had a special plan for those souls and that I am not going to get in the way of that plan. God knew my body and life before He allowed the children to be conceived. I began to dream of a time when I would reunite with those souls that I lost.

In 2014, I felt ready to manifest my son Hunter. I believed that I could conceive and set the New Years' Eve intention of conceiving that year. As friends clinked champagne glasses, I smiled, visualizing that the following year I would have sparkling water in my cup.

I'm not a doctor and I can't begin to understand the myriad of complications that prevent some women from conceiving and carrying children while others find it to be a relatively easy, natural process. However, I do know that the doctor had given me the "all clear" to conceive, and so I used manifestation practices to help my body and spirit prepare.

In the Manifesting Secrets program, you will learn about different ways to elevate your vibration through food. In 2013, I made a decision to indefinitely fast hard liquor because I sensed that it was linked to lowering my vibration and taxing my pancreas, which had already been throttled in my eating disorder days. Alongside a very green, high-nutrient dense diet, I used affirmations daily to tell my body that it was healthy and healed from the years of abuse and bad behaviors.

I took these affirmations a step further by addressing shame that lingered in my body. For instance, I intentionally left in the harsh language above that revealed my inner voice stating, "You kill babies." If the entire foundation of healing is based on love, as I reverse engineer in this book and the Manifesting Secrets course, how, then, could I create life from shame and hatred? I decided to go all in on forgiving myself and releasing the shame that I felt from years of abusing my body.

Earlier in this book you read about the Scale of Consciousness developed over 20 years by Dr. David Hawkins. Dr. Hawkins identified that there's a sort of hierarchy of vibrations in human consciousness. The lowest level of consciousness is shame; it vibrates just above death. Those who live with elevated levels of shame in their lives may feel depressed, compulsively harm others, and even contemplate suicide. Generally, shame is self-directed hatred; it causes you to manifest negative things because your thoughts are so negative; they state that you are unworthy, that you are a mistake, and even that you deserve to die.

Other negative states of human consciousness are guilt, apathy, and grief. Guilt goes hand-in-hand with shame; it's slightly less deteriorating to your spirit and vibration, but it's similarly depressing. You may still have suicidal thoughts when you feel guilt, but the critical difference between guilt and shame is this: Guilt says, "I did a bad thing. Shame says, "I am a bad thing." Guilt often prevents people from forgiving themselves.

Fear, desire, and anger come next on the Scale of Consciousness; but what you'll notice with anger is that it has positive aspects; it can compel you to take inspired action to protect yourself or others, such as the story above about manifesting protection from a child abuser.

Next comes pride, the point at which you begin to feel good. However, pride can enmesh you in beliefs, causing you to see attacks on those beliefs as personal attacks, thereby causing you to feel defensive and closing you off from an open mind and heart to new ideas and beliefs.

Courage is the first level of consciousness that is a true strength. Next, Dr. Hawkins identifies neutrality, willingness, acceptance, reason, love, joy, peace, and enlightenment.

What you'll notice is that as your baseline vibrations elevate higher and higher up this scale of consciousness, you become less attached to the past or the future and more conscious and peaceful within the present moment. In the lower levels of consciousness, you may find yourself withdrawing or preventing yourself from love and relationships due to feelings of unworthiness or fear of being triggered.

When I began to visualize my child and release feelings of shame, I simultaneously tethered my energy to feelings of worthiness. I began to realize that my past was forgiven, that my body was capable of radical self-healing, and I began to peacefully envision my own ability to protect my son from the same abuse that plagued my own life.

Another critical part of manifestation is the feeling of relaxation. When your amygdala is on high alert, your body goes into a stress response state because we perceive a threat. In previous pregnancies, my body was on high alert; I was with an abusive man. As I focused on conceiving Hunter, I got massages weekly, took long luxurious jogs around a lake near where I lived, and planned a life with my future family in a new home in Colorado where I moved in 2014. I vowed to God that if my partner became abusive in any way, that I would take my son to a safe place. As my body began to feel more relaxed, it began to physically heal itself. As we learned earlier, when you are in a state of stress-response or fight-or-flight, your blood literally rushes to your arms and legs, away from your organs. In the parasympathetic nervous state, your blood returns to your organs and you enter a state of rest, digestion, and healing.

When my body sensed that I was financially and physically safe, it relaxed and healed old wounds. The more I relaxed, the more I felt my shame melt away and my heart heal. I believe that this open, healed heart is what allowed me to finally walk into the doctor in July of 2014 and tell her plain as day, "I'm pregnant."

"How do you know?" the doctor asked. "I saw him come to me in a dream. It's a little boy with blond hair." Sure enough, I was only four weeks pregnant at the time I told this to my doctor; earlier than you can generally even identify pregnancy. Hunter is now five and he's thriving; intelligent and healthy as could be. I am so grateful that I manifested my sweet son Hunter.

I kept true to my promise, and when his father and I stopped growing together, we decided to raise our son in separate households and terminate our marriage. I know that it's the best possible choice for my son and he will never know anything different.

CHAPTER TWENTY-TWO:
Manifesting
My Perfect Home

"Confidence is the stuff that turns thoughts into action."
RICHARD PETTY

I've lived in some of the most stunning apartments in the world's most beautiful cities. But when I became a single mom, it wasn't as easy to find a cool flat in the south of France or Hawaii as it was in previous years. I now had another tiny and precious life to consider besides my own, and I hadn't begun to imagine how many factors I'd need to consider when manifesting my perfect home with my son. Unlike my childless days, as a mother I now had to think about available childcare, more space, safety, convenience, geographic proximity to his father, and of course, financial obligations of being stable and self-supporting with a child.

My first apartment in Colorado as a single mom was a disaster; there was mold in the unit, the location was dreadful, the landlord had a fetish for torturing me, and I didn't last a month before I left. Moving with a baby once was brutal. Moving a second time in a month was heart-wrenching and exhausting. For a few years I moved nearly every six months with my son until I finally decided that I needed to employ my manifestation techniques to receive a more stable, long-term situation for myself and my son.

I had decided to live in the center of Aspen, Colorado in June of 2020. At first, my search for a new apartment was looking bleak. I searched high and low in my desired location and, at one point, I even considered living in a studio apartment that was only 300 square feet with a five-year-old. I was that desperate to find a place in my desired location even if it meant selling 90% of my personal belongings to be in the downtown Aspen core. Being that I had become an epic manifester at that time, I took a new approach.

I stopped looking for any place and began to manifest the perfect place.

Here's how that worked:

1. I created a *visualization* of my perfect apartment. In-unit washer-dryer, new floors that were hardwood (or at least not carpeted), two bedrooms, a standing shower, open concept, huge balcony, epic views, and centrally located smack-dab in the center of downtown. It sounded like a tall order, but I believed that a great big God could deliver this dream of mine because of how loved I am, and how much love and devotion I had to others and nature in return. Every day for a month I visualized this apartment; I visualized what it looked like, smelled like, the noises I would hear out my door, and even how I would decorate it. I even visualized the block I wanted to live on, right by my favorite restaurants, museums, and hiking trails.

2. Next, I gathered a *manifestation team* around me. My manifestation team included my entrepreneur mastermind, a manifesting maven friend, a real estate agent who would keep her ear to the ground, my best friend, and even my public following to keep me encouraged. As it turns out, my manifesting maven friend encouraged me to visualize BIGGER things, my real estate agent friend found the perfect place and talked the leasing agent into renting to me, my best friend reminded me to be patient, that the right apartment would come even when it looked like I'd seen *everything* that was out there, and my mastermind encouraged me to negotiate the perfect lease for me.

3. I employed *Confidence*. In my experience, confidence is the most underrated quality of manifestation. Confidence is how we turn thoughts into action, according to expert Richard Petty. Confidence allowed me to approach the contractors, assistants,

leasing agent, landlord, and others to demand what I wanted in the most elegant way possible; despite having a small child at my hip at all times. So often people don't manifest their dreams because they don't have the confidence to ask for what they want.

These three steps put together allowed me to find a truly breathtaking apartment, with many more benefits and amenities than what I expected at the start of my journey. This method of employing a manifestation team, confidence, and visualization with intention and faith has led me to some more epic manifestations that I'll discuss in the remaining stories here in Part Two.

CHAPTER TWENTY-THREE:
Manifesting $25,000 In A Week

"The meaning of everything is the meaning you give it, and your experiences are what you say they are."
David Cameron Gikandi

Normally, manifesting $25,000 in a week isn't a huge deal for me. However, when I was able to manifest $25,000 in a week—right in the middle of 2020 when my son didn't even have school and I was unable to work in my business full time, I felt particularly pleased in my manifestation progress.

One of the things I love about manifesting money is that it doesn't always show up in the form of dollars in a bank account. That is the lesson I will seek to illustrate in this story. So often, we expect that wealth will show up in the form of dollars or material currency. However, after reading this story I want you to consider that manifesting wealth doesn't necessarily mean manifesting money.

In this case, it showed up in the form of:

* Contractors willingly moved to full commission, saving me $4,400 in invoices.

* A contractor refunding me because of mistakes he made on a job; saving me $2,000.

* My new landlord agreed to forego my first month's rent provided I made some renovations on the unit, a blessing of over $3,000.

* My previous landlord refunded me $726 for a week I lived in the home while he was working on it.

* My estate sale made nearly $4,000.

* My apartment was given to me at $600 a month less than the asking price (or a savings of $7,200 for the year).

* A previous client accepted my new rate of $500 an hour for coaching and paid for three upfront sessions ($1,500).

* I lost my credit card and, in doing so, found hundreds of dollars I was wasting on auto-orders I wasn't using each month (approximately $4,800 annual savings).

The total I manifested in one week was $25,826.

Have you been blocking abundance by not showing gratitude for the gifts and savings in your life, expecting all blessings to arrive by check? Be grateful for all your financial blessings and remain open to the prosperity to show up both energetically as well as through trade, gifts, favors, and discounts.

In order to manifest so much money that week, I chose to be grateful for every penny or dollar coming into my hands in any form it did.

Please visit www.manifestingsecrets.com/bonuses for your gratitude affirmations now. You can listen to them each morning while you're brushing your teeth to put you in a vibration for great manifestations that come from a generous God who is eager to deliver our greatest desires, particularly when we show gratitude for what He has already blessed us with.

CHAPTER TWENTY-FOUR:
Manifesting
My Perfect Lover

"Don't settle."

In 2016, my son's father and I divorced, and I felt like my desirability was at an all-time low. I was twice divorced. I was a single mom. I was getting lines, cellulite, wrinkles, and general irritability from aging. I suspected that I'd meet the man of my dreams one day, but I still had doubts to overcome. One day I consulted a dating coach who took me through practices designed to meet my ideal lover. I spent hours sitting alone in nice restaurants and reading books in bars hoping for the right man to suddenly appear. Those practices only made dating a cause for more loneliness and even rejection.

Finally, a friend of mine helped me manifest my perfect lover. She taught me about the Vacuum Law of Prosperity. She showed me that nature loves to fill a void; it doesn't like emptiness. However, with lots of frivolous relationships and flings, I wasn't making myself available to the man I dreamt of.

So, I began to create space for that person. I emptied out the cabinets and cubbies on one side of my bathroom for him. I wrote a list of qualities and put it next to my bed where I'd see it every day. I used specific affirmations and meditations from the Manifesting Secrets program to call in love and romance. I used brain flow yoga to wire my mind for love and healing from any relationships to which I was still corded. I studied with love experts who now teach about love and relationships in the Manifesting Secrets training program. I even began to visualize the names this man would call me, the way he would hold me, the smell of his skin, and the way he moved his body.

When I met Bel, I knew he was the one. We said I love you to each other within 44 hours of the first date, and after that first date I didn't leave his house for days. It was like we always knew one another. This relationship taught me more about myself and about love than any relationship I'd ever had.

Although the domestic partnership portion of this relationship was for a short season, it was one of the most beautiful seasons of my life. I am so grateful that the universe gave me power to manifest epic love and to transmute the pain of a breakup with Bel into the best friendship of my life.

For a free training on Manifesting Epic Love, please visit: www.manifestingsecrets.com/bonuses

CHAPTER TWENTY-FIVE:
Manifesting Healing After Infidelity

"The unconscious purpose of marriage is to finish childhood."
NEIL STRAUSS

There are few things more painful than being cheated on. Furthermore, there are few things that will cramp your manifestation power more than heartbreak. When we persist in the negative vibration of "woe is me" and "victim" mentality after heartbreak, we're attracting more heartache and pain in our lives. They prohibit our ability to learn from the mistakes and take personal responsibility for our own participation in the cheating. Yes, I believe that in almost any conceivable circumstance, you indeed contribute to being cheated on.

I want to start by identifying that I don't believe you're at fault when you're cheated on, although I do believe that playing victim perpetuates the pain. There's a time to accept that you've been the victim of something while still taking personal responsibility for your role. It's a fine line, and I don't mean to downplay the pain of being cheated on. Whether subconsciously or not, we are co-creators in these experiences.

I want to also take a moment to recognize that being cheated on is one of the most painful feelings a person can experience. It hurts and it's good to channel that anger and hurt into personal power, but only after you've grieved honestly. You deserve to cry. For most people who've been cheated on, you experience a sort of twilight zone. If you didn't see it coming, you may feel like your entire world Is shattered, or like the life you thought you had was a lie. It's okay to feel shattered. Once you've grieved, and only after you've given yourself the grace and space to grieve, you can move on to taking that personal responsibility for never letting it happen again to the best of your ability.

In my thirties, I was cheated on for the first time. Months of lying and covering up by my partner left me shattered. I felt ugly, unlovable, foolish, naive, and humiliated. Somehow, I healed, faster than you could imagine, but the lessons trickled in gradually over the years thereafter. With each passing month, I returned to the experience with new, fresh eyes and the lessons kept getting deeper.

In order to heal, I read lots of Esther Perel and Neil Strauss, who pointed out that this and so many other relationship woes I experienced were a direct reflection of my own choices and unhealed childhood wounds. In my situation, I feared being alone and never finding somebody else to love, so I allowed a relationship to perpetuate past its most elegant and graceful course.

There are a few ways that I healed that I'll share with you today.

1. I worked with a life and love coach to help me take personal responsibility for my contribution to the demise of the relationship.

2. My love coaches Brave Legend and Megan Rose Browning helped mediate a conscious uncoupling, the tools from which I still use today, two years later.

3. I forgave my lover and forgave myself for stringing him along.

4. I used the Manifesting Secrets Brain Flow Movement program to open my heart and move the heartache out of my body. Did you know that you can heal your heart by moving your body?

5. I regularly repeated affirmations about my worthiness, specifically focusing my attention on my root chakra and reminding myself daily that I would be able to stand on my own two feet as a single woman and create a new family with my manifestation power.

6. I wrote a new list of all the qualities I wanted in my next epic lover.

7. I visualized those qualities and that man, even hiring a hypnotherapist to create a series of meditations for me to listen to while visualizing my lover's soul and mine meeting.

8. I gave God gratitude for what I learned about love through infidelity; I learned that the person never stopped loving me, and it's for that reason that he cheated; he didn't really want to leave me, but he knew that my love for him was waning.

9. Most importantly, I chose to observe my feelings openly and honestly; I learned to sit with uncomfortable emotions in meditation while breathing slowly and softly; inviting the Holy Spirit to bear my pain while I found solace in silence and surrender.

Healing your heart creates space to manifest more epic things in your life. Tethering yourself to the sadness of a breakup will only attract more heartache in your life. Feeling pain after infidelity is normal. Being honest with that pain is critical. I found self-love as well as compassion for other people after this experience. I realized that the only thing constant in our lives is change, and that change might feel awful; but it always has and always will lead me to bigger and better opportunities. Once I accepted the pain of infidelity and learned to consciously and calmly observe it, I began to release that pain and make it my power.

One of the most common causes for sickness is loneliness. In fact, the primary cause of death in the world is heart disease. Many experts believe that heart-related deaths don't originate from heart defects or disease, but that heart defects and disease are largely symptoms of heartache that festers without healing and, eventually, can kill you.

I invite you to engage in heart-healing with the tools I used. Remember to manifest your next epic love through meditation, movement, affirmations, new language patterns, visualization, story reprogramming, and gratitude.

CHAPTER TWENTY-SIX:
Manifesting Healing After Divorce

*"If you're brave enough to say goodbye,
life will reward you with a new hello."*
PAULO COELHO

A life coach Brave Legend once shared with me, "the wounds of a relationship are best healed in a relationship." For this reason, I didn't pressure myself to fully understand my divorces and the role I played before moving into another relationship where my own issues were front and center, once again. However, I made epic strides in understanding, healing, and forgiving myself and my former husbands with the tools in this story.

I love to manifest through language. For instance, I began to refer to my son's dad as my "co parent" instead of calling him my "ex." Your language is one of your most powerful tools for manifestation; your thoughts literally create your reality. Your words are blessings... or they are black magic. In addition, the words we speak manifest our feelings. If I constantly use the term "ex" to describe my previous partner, I was saying that there was a big void or failure in my past. For that reason, I began to call Hunter's dad my co-parent instead of my "ex."

Another tool I used were brain-rewiring tools to replace hatred with love. One easy way I did this was to memorize and daily repeat quotes that helped me find compassion for myself and my former partners. The primary quotes I used were Dr. Martin Luther King Jr.'s "Let no man pull you so low as to hate him" and "Darkness does not drive out darkness, only light can do that. Hate does not drive out hate, only love can do that."

In addition to manifesting love where there was anger and resentment, I learned to see my former husbands as a parent sees a child; I began to see him through God's loving and forgiving eyes as somebody who is trying to do his best. I could no longer fault him if he was doing his best, and in the same way I began to forgive myself.

Another powerful way I manifested love in place of hate was to keep images of my former partner on my altar so that I would be forced to look at a face that made me cringe and learn to see him as a child. In fact, I even put the photo up next to one of myself as a small child. I thank both of us for doing our best. Seeing us as children began to give me grace. Whenever I see the photo I attempt to stop and say the Ho'OPonoPono prayer which is simply, "I'm sorry. Forgive me. Thank you. I love you."

I began to incorporate love into my meditations, taking time each morning to breathe love into my heart and radiate it toward the image of my former partner. I began to visualize his own heart softening and strengthening as I sent that love in his direction. As with any lesson, we often revisit things we "think" we've learned so that we can learn them on an ever-deeper level. In 2020 my co-parent and I got into a scuffle that caused me to revisit the above steps once again with a vengeance; I had much more pain and anger to overcome and transmute into the manifesting power of love.

Another tool I used to manifest healing after my divorce was to use the emotional and brain regulation tools you'll find in the Manifesting Secrets training. Among these are breathing exercises designed to regulate your nervous system, turn off your "fight-or-flight" response and turn on your parasympathetic nervous system, and help you feel almost instantaneous peace and calm.

Divorce has been a remarkable teacher. It has taught me to not take anything personally. Having been divorced twice, I've learned to transform my anger into compassion. Both my romantic and platonic relationships since my 2016 divorce have been remarkably deeper, intimate, and honest.

Divorce also helped me to forgive myself for my own anger. I recognized that each time I felt anger or hatred, my inner child was begging to be nurtured, held, supported, and told she's safe. I took time to share out loud with a trusted friend that I felt scared, hurt, and angry. With this honesty and vulnerability, conversations opened up to deep, constructive strategies for healing the new wounds with tools I'd learned from my first heartache with divorce.

Finally, I committed to speaking positively of my former partners, even among my best friends with whom I would otherwise be "safe" to speak negatively. When I chose to speak highly of former partners, I manifested healing for them, for me, and even for my son. I regularly heard stories of how one man chastised or gossiped about me, but I committed to taking the high road. In doing so, I align myself with high frequencies of love and light versus black magic and hatred. In doing so, I began to receive more love and light from the world around me. What we feel we attract, and healing from my divorce helped me to manifest a brand-new life independent of the scarlet letter I had originally worn from childhood wounds and indoctrination.

CHAPTER TWENTY-SEVEN:
Manifesting Epic Love After Divorce

"Your new lover will thank you for having the courage to leave a situation that wasn't right for you and count their blessings that you became available to find them."

In retrospect, I thought that the very act of divorce would break the subconscious patterns and self-sabotage that ended my marriage.

When the same patterns that seem to destroy my marriage begin to show up with new, more incredible, more loving, and altogether categorically different types of partners, I knew that the problem was not the partner, but me.

Although I believed that, "the wounds of relationship are best healed in relationship," upon sabotaging my first epic love post-divorce, I struggled and still periodically see glimpses of the fear that I'm getting too old and nobody wants a single mother living in the middle of nowhere. Even in writing it, I can't believe that such shallow, superficial thoughts could enter my magnificent mind.

I am capable and worthy of receiving the perfect partner at the perfect time.

I've spent most of the four years since my divorce single; diving deeply into my codependent, avoidant, and downright persistent patterns of avoiding love and the magical vulnerability and refinement that it brings.

One of my all-time greatest teachers has been the one partner I have had in the last four years, our ever-evolving friendship and the way he serves as a mirror for my refinement.

Ultimately, if I had to identify one most profound difficulty in dating after marriage (for me), I would say it's the victim mentality: my tendency to blame everybody else instead of taking personal responsibility for my life and love.

Happily, I have been remarkably effective in recent months at crucifying this toxic tendency thanks to lots of visualization, daily meditations and mantras, humility, and being downright sick of my old patterns.

Most especially, the pivot is thanks to 2020; a year that destroyed the spirits of so many single parents who were forcibly removed from the workforce with a modicum of the time we previously had to earn money and experience pleasure the way we used to.

Not only have I re-programmed my brain to believe that everything always works out for me, but also that every obstacle is truly an opportunity. And with my mind, I am absolutely capable of creating ineffable happiness and joy no matter what the circumstances around me.

CHAPTER TWENTY-EIGHT:
Manifesting
My Dream Job

"Work never goes away,
but when you love what you do,
even work is pleasurable."

I wasn't passionate about any specific job or profession as a child. Even in my thirties when I had my son, I frankly had no idea what my life's purpose was. I did, however, know deep down that I had one. That's when I began to read every book I could get my hands on to learn about finding purpose. Over the next four years, I came to a very solid conclusion that my life's purpose was to inspire people. I learned that I had an uncanny way of spotting and illustrating other people's genius. I could make people feel superhuman by simply reflecting what I saw in them.

MY FIRST COLORADO
LISENCE PLATE

1NSP1RE

MANIFESTING
SECRETS

I took this new-found discovery of myself and created a series of programs on manifestation and confidence. I even began to help others create and publish their own books and programs because I had such adamant belief in those students. I developed a seven-step process that takes people through the questions they will need to ask themselves and the worksheets needed to dig into their subconscious programs and sometimes hidden desires. The steps include:

1. Rewrite your past stories
2. Find your flow state
3. Identify your zone of genius
4. Attach yourself to what feels good
5. Find a way to connect with and contribute to others
6. Visualize your dream day, every day
7. Sit in silence and surrender to signs and synchronicities as they arise

One of the best things you can do to find your life's purpose is step two. Imagine a time when you felt like you were in a tremendously productive state; like you were downloading wisdom directly from the Creator. This is a sign that you are in your purpose. For me, I identified that almost anywhere I went I could find a way to inspire people; sometimes I would even bring a barista to tears with my encouraging words. I noted that most of my clients were able to quit their jobs and follow their own passions full time. I recounted how many friends of mine were filled with confidence because of the life that I breathed into them.

Alas, I found that my life's purpose was to inspire people. It's even on my license plate. I spent time over the next year on camera as much as humanly possible to inspire people. During this year my brand and businesses exploded; once I knew my life's purpose there was no end to the prosperity that I received. If you feel unfulfilled or apathetic about your life's work, these emotions are among the lowest on the Scale of Consciousness we discussed in an earlier story. Conversely, the feeling of satisfaction and flow state that comes with doing something you love will manifest not just happiness, but more epic wealth.

When we courageously use the Manifesting Secrets tools to improve relationships, cognitive function, and language patterns, we draw closer to our dream careers as the universe puts the people, projects, and plans into our laps. When you speak confidently and with faith about what you want and who you want to serve, you magnetically attract opportunities to do so into your life.

Part 3

7 Steps to Start Manifesting Today

"In neuroplasticity, habits help us to form the circuitry to make manifestation easy. Neurons that fire together wire together is a concept in neuroplasticity that illustrates how what habits we form will become subconscious, automatic, and effortless; thereby making manifestation easy. What we don't use we lose; this concept illustrates how we can rid our lives of those things we don't want to carry with us anymore by simply focusing on the words, thoughts, beliefs, intentions, and practices that will replace previous behaviors, beliefs, programs, and otherwise indoctrination."

The Manifesting Secrets Program details 30 practices that you will use over the course of 90 days to become a more effortless manifester. Some of these practices are quite simple, such as giving yourself permission to heal previous patterns that no longer serve you. Some are quite complex, such as aligning chakras or creating visualizations designed to help you manifest.

In Part Three of *The Manifesting Mind*, we're going to discuss the seven easiest and most fundamental ways to manifest. Some of these you may be familiar with, but you will never have heard them taught in this format before. To enjoy the accountability and power of our 90-day program and the remaining 50+ manifestation exercises, simply visit us at www.manifestingsecrets.com.

If you want a free video on manifestation through neuroplasticity, please visit www.manifestingsecrets.com/bonuses to download yours right now.

Reprogramming Thoughts with Mind-Rewiring Language

"A man's mind may be likened to a garden, which may be intelligently cultivated or allowed to run wild; but whether cultivated or neglected, it must, and will, bring forth. If no useful seeds are put into it, then an abundance of useless weed seeds will fall therein and will continue to produce their kind."

JAMES ALLEN

Now that you are familiar with how Manifesting Secrets works and a few examples of how I've applied it to my life, here are a few practical steps you can take starting today to begin consciously choosing the life you want to live once and for all.

The first step in Manifesting Secrets is through language for manifestation. It's not enough to say affirmations or mantras, but it's important to use these language patterns in a practice with intention, such as while looking in the mirror or while practicing your breathwork or Brain Flow movements from the Manifesting Secrets program. Try using yoga movements to carve your affirmations and intentions into your mind. In fact, I have my students listen to recorded affirmations every morning to help each day flow with more alignment and connection to the Source.

What is an affirmation?

In short, an affirmation is something you say over and over again until you believe it. Essentially, affirmations are designed to help you alter the way your brain thinks. They are designed to reprogram your mind to believe the positive affirmations you use. Affirmations are one of many powerful practices we use to help with rewiring the mind. By combining the other six steps in the Manifesting Secrets seven-step process, you can rewire your thoughts more quickly.

By now you have learned that manifestation is the act of materializing your thoughts. Your thoughts may be negative, hateful, or even poor or impoverished. And, too, your thoughts may be abundant, positive, loving, and generous.

You understand that you are always manifesting. You are always creating your reality based on thoughts and subsequent actions you take from those thoughts. However, you can't rewrite an entire life and mind of negative thoughts by the end of this book.

One of the best things to do when you find yourself in a negative situation or thought pattern is to stop and find out where that thought is coming from. Is it coming from past trauma? From fear for the future? Is it coming from subconscious patterns you inherited from a negative caregiver or family member?

Once you isolate the core thought that causes a cascade of negative thoughts with it, you can manifest more quickly.

It's very rare that people become instantly happy, healthy, and wealthy overnight. This process takes time. In fact, scientists agree that it takes between 8 and 20 weeks to rewire one single thought. That is, it takes between 60-140 days to make a new thought automatic, or "subconscious."

I'll give you an example. Earlier this year I was having confrontations with my new neighbors. The guy downstairs sued my landlord because of toddler noises that were annoying his tenant. The neighbor down the hall sued the same landlord because a contractor supposedly removed asbestos from the unit without a permit. It was a non stop barrage of lawsuits and people nit-picking at one another in harmful, painful ways.

During that time, I created a tragic series of legal matters in my own life; a family member started filing non-stop motions, modifications, and other suits against me for innocuous things and even several false accusations.

I knew I had two choices: complain and play victim or show love to the persecutors, while minding my business, and remembering that everything always works out for me. In time, the suits were settled, the modifications were made, and life returned to peace in Aspen at my apartment. While the litigation with my family members persists, I remember every day that "everything always works out for me." At this point, I've said it so many hundreds of times that I've finally started to believe it. Not only has my faith increased, but my peace has increased. I don't have to try as hard to attach myself to feelings of bliss and contentment; those thoughts of peace and comfort now come automatically to me.

You've heard of affirmations. But what you haven't heard is that rewiring your thoughts goes far beyond saying positive things. It's a practice of repetition every single day until your subconscious mind has created a new groove wherein that thought becomes automatic.

A pivotal moment for people who manifest positive, powerful, prosperous things is when you begin to make the subconscious conscious. As you're about to learn, your subconscious thoughts are underlying beliefs, behaviors, habits, triggers, or traumas that you don't usually consciously recognize.

When you control your thoughts from negative to positive, you literally create Heaven in your mind.

Affirmations are the tip of the iceberg to manifestation. Books like *The Secrets* claim that affirmations and visualization will attract your material desires.

However, while I do believe in "instant manifestation," I understand that permanent reprogramming of the mind takes repetition, movement, meditation, and more practices outlined in the following pages.

Although affirmations aren't the final answer in manifestation, using affirmations properly can have a measurable impact in your life right this moment. That's because affirmations are a discussion with the Universe. What you speak reverberates throughout your entire body and life.

When you say something positive, you attract more positivity in your life. It resembles The Observer Effect, a concept in quantum physics that can be best summarized as, "what you focus on expands."

Similarly, negative thoughts, self-speak, or words cause negativity to expand in your life.

As you begin to use the mind-rewiring affirmations I provide in the back of this book, you will begin to believe and manifest the moods and material abundance in the phrases; you will convince your mind to believe them.

It doesn't happen overnight; give yourself time and be very patient with yourself.

Practicing Step One

Choose some of the affirmation sheets at the end of this book to tear out of your book or to print from www.manifestingsecrets.com/bonuses

By repeating these affirmations daily, you will begin to enact the Law of Attraction; you will begin to feel good things when you read these positive words. Elevating your vibration to faith, hope, and love will help you attract more good things into your life under this simple principle: like attracts like.

To understand this better, imagine a time when you walked out of the house in a phenomenal mood. Your boss recognized your positivity, your kids were happier, even the server at dinner sent you a special drink on the house for brightening his day. When we live with high vibrations of love, joy, peace, and compassion, we infuse light to the world around us and that light reflects back to us more of the high vibrations we emit. Your light lights the world around you and, in turn, is reflected back to you by the people whom you shine this light upon. If you doubt this, try it and you'll see almost instant effects of being light in your world.

The second part of this exercise is to choose just one affirmation that you will be saying every single day. Take a Dry Erase marker and put it on your mirrors; write it on your fridge; put it in a note on your car where you will see it instead of your clock.

Once you say this affirmation enough times with conviction, your brain will begin to believe it. Remember to give yourself eight weeks or more to rewire one thought; and allow yourself to take up to twenty weeks if the thought needs extra space to solidify in your mind programs.

Take a moment to write your first mind-rewiring affirmation here:

- -

- -

- -

- -

- -

- -

- -

Download your own recorded affirmations by visiting www.manifestingsecrets.com/bonuses today.

Leaning into Manifestation with Intentional Relaxation

"Here there remains nothing to do or achieve.
You have entered the domain of grace.
All that can never be done by your doing can happen in
the non-doing presence of your being."
Dr. Kamini Desai

The best place to practice manifestation is in relaxation. Whether this is while meditating, saying affirmations, or while monitoring your self-talk, relaxation is key to manifestation. But trust me, that isn't easy for me to say, and much less easy for me to practice.

My tendency is to hustle. I'm your quintessential oldest child, type A personality, overachiever, manic perfectionist. But today I'm going to urge you to stop hustling because the magic is in alignment. It's been one of the greatest ways I've learned to manifest in my own life, in fact. Manifestation and living in relaxation necessarily mean working less hard and working with more inspiration, flow, steadiness, and even slower pace.

When you relax into manifestation, you move with more flow and faith; there is less chaos and frenzy clouding your vision. You begin to do less and receive more.

When we relax, our brains are propelled into more eager states of neuroplasticity and learning, as my many teachers and the neuroscientists who have contributed to the Manifesting Secrets program have illustrated. When we sleep restfully (not fitfully) or relax, we detoxify our bodies as well our brains. When we are awake and still in this relaxed (not apathetic, but peaceful) state, we experience more self-healing, clarity, confidence, and wisdom in our days, which materializes into a better life.

During relaxation, our brains can focus on choosing our thoughts through the process of neural differentiation, the "this not that" process that you need in order to strategically choose positive or negative thoughts (which beget a positive or negative outcome in healing and life).

Relaxation allows us to remove the obstacles to healing that arise from simply being too "busy." When we're focusing on tasks, we aren't identifying the solutions that remove obstacles in our way. Fight, flight, and freeze modes create an environment that makes it impossible to learn. Relaxation is a place where we can remove fear in order to learn and remap our brains.

Fear shuts down our amygdala's response and sympathetic nervous systems. A teacher I mentioned earlier, Lauryn Gepfert, refers to the state of relaxation as KPE, or a place of kindness, patience, and encouragement.

Here's a short summary of KPE:

* You must be kind to yourself in order to feel relaxed enough to learn.
* You must give yourself space to take as long as you need to learn a new pattern.
* You must tell yourself that if it takes 10,000 times, you can take as long as you need to learn, grow, and heal.
* You must encourage yourself along the way—it's okay when you mess up; you've got this.

Fight, flight, or freeze mode causes an alarm in our brain's amygdala; this makes it impossible to learn; KPE puts us into a parasympathetic state of learning, growing, and healing. Your brain is in a state of neuroplasticity and learning when you are relaxed; feeling tension, struggle, or anxiety will alert your brain to a threat and prevent you from growing.

**THE BEST PLACE FROM WHICH TO
LEARN, GROW, AND HEAL**

K	**KINDNESS**
P	**PATIENCE**
E	**ENCOURAGEMENT**

MANIFESTING
SECRETS

These are "safe" brain states where fear is removed and love leads. In fact, KPE are the exclusive parameters within which learning must happen, according to Gepfert. If you are afraid, you won't get something after the first "affirmation" or conscious strategic "gaining" of a thought, give yourself permission to practice 10,000 times.

We learned that in studies done of Tibetan monks and monastic priests, it has been found that a relaxed, meditative state of being causes the brain to fire on all cylinders. Perhaps this is why many believe the world's happiest person is a monk.

In short, reject the rules that you learned about doing and embrace the joy of being. Reject the thoughts that it has to be hard to be good. Reject the thought that if it doesn't happen quickly, then you aren't doing it right. Reject that idea that you're doing it wrong at all. Relax into KPE.

Rejecting the rules or the old, negative thought patterns actually has a neurological function.

The science around this is called mono-directional healing.

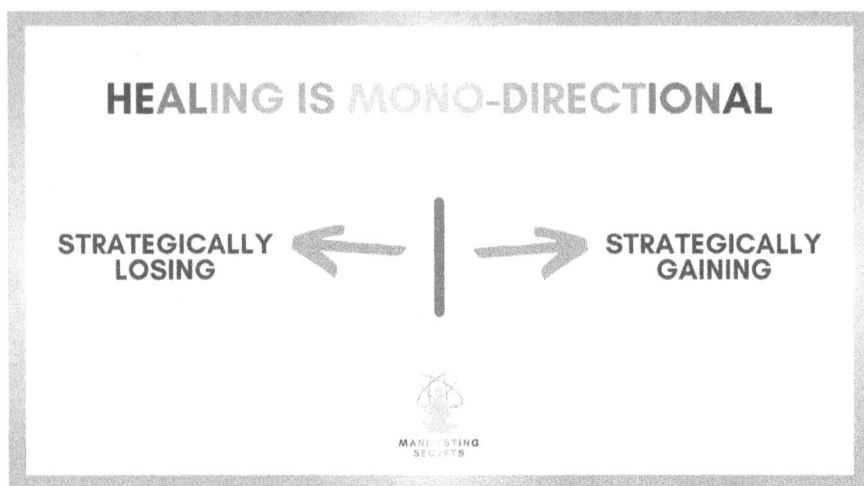

You see, when you want to heal, you must choose to strategically lose in order to strategically gain. Choose to reject the old thoughts and then set an intention for what you want. Strategically reject what you don't want by "overriding" your previous brain patterns through your Manifesting Secrets training.

A practice I like to do every day on the hour is to step outside of myself through books on tape or mindfulness practices. When I feel a negative emotion, I reverse engineer that emotion right down to the thought.

Perhaps the thought is one of criticism of another person. For instance, a former partner of mine recently spent time with a woman who didn't support me during our separation. I created stories in my head about her worth, her looks, and even her family. I listened to myself criticize this woman and chose to reverse engineer the thoughts I had: "I called her this name because I feel the same about myself. I judged her worth because I feel unworthy. I assumed she doesn't like me because I feel self-doubt and self-loathing around my separation." I choose to reverse engineer right down to the past traumas that led me to said thought. "I criticize her looks because I am self-conscious and grew up feeling ugly and fat."

By this process of reverse engineering my own criticism and hatred for others, I can strategically lose my feelings of being offended, hurt, or judging others. In place of those negative thoughts, I gain the following thoughts mindfully: "Julie is a nice woman. She is a great mom. She is beautiful. She has a voice like an angel. She loves my former partner and is a great influence on my son. I love Julie."

In the spirit of kindness, patience and encouragement, I learn to love my "enemies" so that I can strategically lose thoughts that seem to be critical of others but are really judgments of my own self and projections of my own pain and trauma. I choose to then strategically gain thoughts of love, acceptance, non-attachment, and peace.

This allows me to free up brain space spent in negativity and channel it for manifestation of what I really want, which is love, health, wealth, and happiness. I reroute the negative emotions into positive ones and then get back to the momentum of love, high vibration thoughts, positive emotions, and acceptance in my world and the people in it. By strategically losing the old negative thoughts, I strategically gain not only peace, but manifestation power in my life.

Please learn about how Lauryn Gepfert is healing people with the power of their minds at www.nfiheals.com

Elevated Cognition

Choosing your thoughts wisely doesn't just help you rewire negative patterns that keep you stuck, in scarcity, and sad... but negative thoughts are critical to seize because they are believed by Manifesting Secrets experts to cause disease in your body as well as dis-ease in your psychology.

One exercise that Dr. Christiane Northrup uses to help people seize and reprogram negative thoughts is through Elevated Cognition. Elevated cognition is the process of choosing a thought that feels best in real-time.

Let's imagine you're looking to close a new client sale, apply for a new job, or otherwise sell yourself. Suppose you think to yourself, "Nobody will want me. My rates are too high. I'm not smart enough for this gig. Plenty of people can do this better than me."

Somewhere in your life, these thoughts developed as a response to a trauma or due to your training before age seven. The negativity is the result of subconscious programs. Replacing those negative thoughts with more positive, higher vibration thoughts is critical to remapping your mind.

You can say, "why wouldn't they want me? I'm highly competent. I love myself and this person will love me too. I'm worth my rates and they'd be lucky to have me. If I don't take this client, another better one will come along..." As you learned earlier, biologist Dr. Bruce Lipton believes that up to 95% of our thoughts come from subconscious programming. Scientists have also found that we have over 60,000 thoughts each day; and the majority of these thoughts are negative.

Practicing Step Two

Today we're going to practice elevated cognition as your first exercise in KPE, or Kindness Patience and Encouragement, as a way of living.

First think of something you want. It could be a relationship, more money, or even better health. Now imagine why you can't have that thing. Perhaps you think you're unattractive or unintelligent. As you become a powerful manifestor, you will listen to your inner dialogue very closely.

This process of "hearing" your own thoughts as a conscious observer takes practice. Over the next few weeks and months, you will begin to catch yourself thinking negative thoughts in real time, and you will be able to rewire them as well. This won't happen overnight; it takes lots of consistency, repetition, and of course, KPE.

Creating new patterns in your mind literally means that your new thoughts, when repeated over and over, become hardwired; you create a new groove with positive thoughts where the negative grooves used to be. Perhaps you don't believe you're worthy of a great relationship. However, as you repeat the positive thought in place of that, such as, "I'm a great catch. I'm worthy of love. I am love," you will begin to believe these thoughts.

In some ways, neuroplasticity is like making your brain your slave. Dr. Northrup encourages you to take this task a step further by arguing with your subconscious in a playful way. "You think you're ugly? You're amazing; how clever of you to think that."

When we detach from the emotions that are behind a thought and focus on appreciation for every thought without judgment, we're engaging in the practice of Kindness, Patience, and Encouragement. This process of seizing thoughts and infusing even the negative ones with humor was detailed on the Hay House radio show by Dr. Northrup. She emphasized the need for humor in your neuroplastic activities because humor is, as she says, an "exalted emotion." Every person has a shadow side. The shadow is the part of you that you are likely inclined to hide, repress, or deny.

Hiding, repressing, or denying your negative thoughts is actually counterproductive. One may think that repressing a negative thought is a positive act, but it ricochets in a forceful way to compound negativity.

You see, when you resist something, it persists. I believe in part it does two things: it hinders our ability to have clear thinking because it puts us into fight-or-flight and, second, it draws attention to what you *don't* want.

You aggravate or emphasize your negative thoughts when you repress them. This brings us to step two in Elevated Cognition.

First, consider what you take in from mass media, magazines, television, radio, and the people around you. I personally carefully curate what I see from "mass media." I carefully curate what comes into my mind. I do not own a television and I rarely watch movies or even Netflix.

You see, every sound and image we consume has a vibration. We learned about that in Module Four while discussing energy consciousness. David C. McClellan of Harvard University did a fascinating study to confirm this concept. In this study, medical students were shown two different versions of a movie: one positive and another negative.

During the study, McClellan measured levels of IGH in the students. Levels of IGA in our bodies correspond with our immune system. Low levels of IGA are a sign of weakened immunity, and vice versa.

The medical students who were shown a positive movie about Mother Teresa experienced a rise in IGA. However, IGA plummeted during a video about war. Darkness feeds on darkness, hate feeds on hate; it reminds me of a famous quote by Dr. Martin Luther King Jr. that I referenced in my above story about Healing from Divorce.

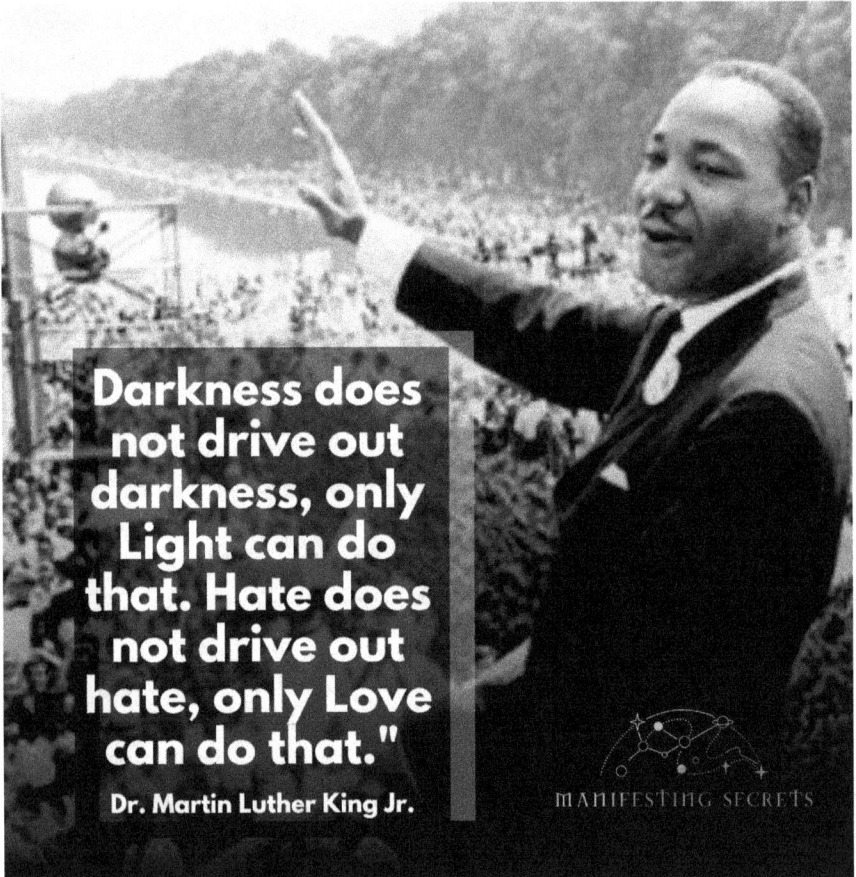

Darkness does not drive out darkness, only Light can do that. Hate does not drive out hate, only Love can do that."

Dr. Martin Luther King Jr.

MANIFESTING SECRETS

Conversely, however, love feeds on love. Love creates more love. In my life, choosing to love exponentially affects the world around me for good, whereas hate has little power except to cause me to harm my own self. Love, in fact, affects the vibration of the world exponentially more than hate, as we learned in our study of Dr. David Hawkins' Scale of Consciousness.

Dr. David Hawkins asserts that a person vibrating in a compassionate frequency actually counterbalances thousands, if not millions of people vibrating at a negative vibration.

Answer the following questions in your book or on a separate sheet of paper.

What is one thing you want more than anything else today?

- -

- -

- -

- -

- -

- -

Practice saying this mantra ten times in a row very mindfully in front of a mirror while looking into your own eyes. Start by saying it when you wake up, then set your cell phone alarm or a mindfulness app to say it hourly.

Dear (your name), you want

- -

- -

- -

- -

- -

- -

You will make this happen. Even if it takes 10,000 tries or 10 tries; you will get it done. You are fully equipped, capable, creative, and worthy of this achievement.

Thank you for being so fully equipped, capable, creative, and worthy of this goal. You trust in the Universe to send the people, projects, programs, teachers, mentors, resources, funds, strength, and courage to make this dream a reality.

STEP THREE:
Meditation

*"Meditation is not just blissing out under a mango tree,
but it completely changes your brain."*
MATTHIEU RICARD

Like so many others in the Western world, my meditation practice wasn't delivered to me via family modeling. Rather, I began meditating when I heard a radio ad for a meditation app.

I was a new mom, and I was frazzled, exhausted, tired, and feeling hopeless much of the time.

I was promised that ten minutes of meditation daily would help me to clear my head and become a more conscious observer rather than emotionally charged about my life and family. I heard the analogy of life being like a storm. I could be in it or observe it. Heavens, did I want the latter? So, I dipped my toe into a meditation challenge.

I began noticing more happiness, clarity, and calmness in my days. Sometimes I would meditate when my son would fall asleep in a parking lot. I was an eager student and needed to steal any spare moment I could find to meditate. Most importantly, I committed to finding that time despite being a single full-time mother with an infant. I made it happen.

Meditation is said by some to be the modern foundation of self-care. However, it's been around for millennia. Whether you're seeking enlightenment or just need to feel less stress and anxiety, focusing inward is the path. Before you buckle down into some deep breathing, let's discuss how meditation makes a measurable impact on your brain. Meditation helps us focus and clear our minds.

Meditation and Brain Waves

I enjoy falling into meditation first by breathing deeply and then noticing any noises around me. Invariably, those noises fade as I go deeper into my meditation. As you make these observations while falling into meditation, your brain shifts into more alpha and theta rather than beta brain waves; meaning that you begin to slow down as you practice meditation.

When I first began incorporating deep breathing and meditation into my life, I used to meditate while my infant son was crying. During times when there's nothing I could do to soothe his colic, I would hold him and stare into the darkness, gently rocking him and breathing deeply.

I would envision his own breath slowing down from choppy, staccato sobs into my own slow, sweet rhythms just as my own choppy beta brain waves slowed down into more alpha states.

The deep concentration you associate with meditation doesn't usually happen when you're new to meditation; but as you practice, you can develop these states through stronger "gamma" waves you enter in meditation.

How The Brain Lights Up When You Meditate

When you meditate, your brain doesn't "fall asleep." In fact, in studies referenced earlier of Tibetan monks and Monastic Priests, scientists found that while meditating, your brain actually lights up. It is said to increase your ability to heal your body as well as rewire negative thought patterns in your mind. Meditation is a tool used by teachers of neuroplasticity; or the brain's ability to rewire itself to help you learn, heal and grow. Incorporating meditation into your life helps your brain become more plastic, or able to change its wiring and firing patterns. This means that you will have the opportunity to literally heal your body, manifest more prosperity, and even call more love into your life.

So much of what prevents us from love, prosperity, and healing is actually due to negative thought patterns as well as the "fight-or-flight" state that we find ourselves in too often in our demanding modern world. During meditation, your pre-motor cortex and insular cortices increase their activities, which is why scientists found that the brains of meditation priests and monks were so neuroplastic. Quieting your mind, alas, ignites your brain. Your pre-motor and insular cortices help you regulate your attention and give you awareness of your body, thoughts and emotions.

Meditation is a muscle; it grows stronger as you train it. You may struggle to focus during meditation when you first start, but you may find yourself more readily falling into a highly concentrated, focused, meditative and state of awareness as you practice meditating.

Losing Focus is Not the Obstacle, But the Way

This is one of the most significant findings of my research, so I want you to really pay attention to what you're about to learn if you have any desire to meditate or self-limiting beliefs around it.

First off, I'm crediting the title of this section to Ryan Holiday, author of *The Obstacle is The Way,* one of the best books ever written, in my very knowledgeable opinion. Ryan, you're a hero of mine (I know he's going to read this book).

Meditation is believed by many doctors and healers to connect us to the quantum field. What I mean by this is that mediation helps us to connect with our future selves; it helps us supersede and transcend material linear timelines and gather strength from a future, more enlightened self. This is a lot of where faith comes into manifestation. When you connect with your future self, you have more faith in him or her. You have more confidence that you will materialize what you manifest today.

In this field I call the quantum field, fear doesn't exist, love is infinite, time is relative, resources are limitless, and our brains relax firmly into the present moment. This is the place where our brains light up and we heal wounds and traumas that have been blocking us from advancing in our enlightenment or manifestation. As mentioned, there is no possibility of learning, healing or growing while we're in a state of stress, tension, or fear.

So why doesn't everybody meditate all the time? Sometimes, meditation feels like another "to-do." We often criticize ourselves in meditation because we get distracted.

Here's the good news: one study by Emory University indicates that meditation isn't the most effective by simply creating a goal of reaching transcendental states. Indeed, this stands in direct contradiction with what so many of us have been taught is the goal of meditation. The goal isn't the transcendental state, it's the shift of attention.

Here's how that looks: Through advanced FMRI machines, we have learned that the brain is best served in meditation with this four-part formula.

1. *Focus*

2. *Mind Wandering*

3. *Awareness*

4. *Shift*

If you lose focus (step two) 1,000 times in your meditation, you will be best served not to condemn yourself or tell yourself you aren't doing it right, but to thank yourself for your awareness. Congratulate yourself for your efforts. And then shift the mind wandering back to focus.

According to this Emory University study and so many others since, it is in the shift back to focus from mind-wandering that produces the most profound changes in your brain for the neuroplasticity we seek in manifesting a new mind and, therefore, new life.

Meditation to Relieve Stress and Manifest More...

If you're using meditation to manifest a better life, body, or relationship, you're wise to head in that direction. Indeed, meditation can help us almost instantly move from fight-or-flight mode (our sympathetic nervous state) to the parasympathetic nervous state, or what my friend Rowdy, a stress and anxiety relief coach, calls, "Rest and Digest with Ease and Flow." If your goal is to manifest learning, growth, and healing through meditation, you won't get there when you're in your sympathetic nervous state or this fight-or-flight state. That's because when you feel stress, tension, fear, anxiety, or a threat, you're designed to focus on that threat. However, healing, growth, and learning can't take place when you're in fight-or-flight. In fact, it's neurologically impossible.

If you want to manifest a better life, relaxing and returning from fight-or-flight states is critical. If you're having trouble sleeping or emotionally regulating, you may be stuck in fight-or-flight mode.

Once you begin to meditate, you will find that a fifteen-minute trip inside your own head and heart can be more rejuvenating than a day at the spa. Indeed, many experienced meditators feel addicted to the practice; and if there's any practice I condone an addiction to, it's meditation. I, for one, am quite addicted. As a former sufferer of anxiety and depression, I am pleased to know that medical experts across the board agree that, without a doubt, meditation can even help with symptoms of anxiety or depression.

Meditation Is the New Black?

Although meditation is having a sort of pop-culture resurgence, it's been around for millennia. Scientists are having great success scrutinizing the links between meditation and optimal brain function for healing, learning, and growing. Schools are beginning to replace detention with meditation to help kids feel less anxiety or aggression.

Offices are hosting meditation classes and meditation break rooms for employees who want to recharge, refocus, and even interact more positively with the rest of their teams. It's a win-win for everybody when a team or company is operating from a more neuroplastic state; they think more strategically with ease and flow; they operate with less stress and clearer heads.

According to medical professionals, stress causes between 60-90% of our non-vital functions to shut down. Meditation is paramount to the work these scientists and healers do for patients looking to rewire their brains even so powerfully as to bring movement to paralyzed limbs, like the teachers I mention in my introduction.

Meditation helps contribute to increased attention span, improved memory, quieting the amygdala, and more activity in your prefrontal cortex. Many studies also indicate that meditation actually increases cortical thickness and the density of gray matter. What's more, it is shown to increase coherence between both hemispheres of your brain.

A study conducted by Chiesa, Calati, and Serretti suggests that you only need eight weeks of consistent meditation to improve cognitive function. In fact, according to studies, as soon as the very first time you meditate, you feel more connected with and empathetic toward others as your brain balances your ventromedial prefrontal cortex and the dorsomedial prefrontal cortex.

Meditation can immediately help you return to a state of rest, digest, ease, and flow because your fear response subsides via the amygdala and the two branches of the nervous system. From the first time you meditate, you begin to create new neural pathways for manifestation to take place. If you're feeling anxiety or stress, use meditation to reduce cortisol and give you more control over your life.

Your amygdala initiates stress response reactions before your brain even knows what's going on. It tries to protect you from experiencing pain, danger, or loss. Mindfulness and meditation allow you to observe stressors with curiosity, compassion, and non-judgment. You will know when your body is preparing to fight or flee when your breath becomes shallow or short. That's why the best way to start any meditation is by slowing your breath.

Controlling your stress response is key not only to emotional regulation, but also for long-term health and neuroplasticity. Cortisol, which increases during stress, actually damages all tissues in your body, including cells and neurons. Meditation decreases your threat response and anxiety, helping you respond more mindfully to the world around you and lowering cortisol for a period during and after you meditate.

Do you feel nervous or anxious throughout your day? Do you feel fear? You could be causing the toxic hormone cortisol to flood your body multiple times a day, or even per hour. Could the cessation and even reversal of cell damage or neural damage be obtained through meditation? According to science, absolutely.

Activating "Bliss Brain" Through Meditation

One of the most powerful books you'll ever read on meditation is the book *Bliss Brain* by neuroscientist Dr. Dawson Church. We have referenced it in the Science of Happiness Chapter. Dr. Church identifies therein a powerful practice he calls Eco Meditation. Here's the short description of that meditation in case you would like to try it in your own practice.

Tap to release any stress in your body or emotional obstacles to inner peace. You can gently tap the outside of your hand between your wrist and pinky finger 5-7 times. Tap above your eyebrow, between the eyes, below the eye, above the mouth, below your bottom lip, in the center of your clavicle, under your arm, and then back to your hand.

1. Relax your tongue on the floor of your mouth to activate the parasympathetic nervous system.

2. Slow your breathing down and breathe through your heart.

3. Imagine a beautiful place with people you love.

4. Send a beam of heart energy.

5. Picture a big empty space behind your eyes.

6. Envision the peace you feel right now and imagine that you can tap into this same peace at any time, in any place. Send gratitude to yourself for your practice today.

7. On that note, remember that meditation is called a practice for a reason. Like any new habit, the first few dozen times of practicing take more effort than after you've developed the meditation muscle, so to speak.

By the time you read this book, I am manifesting that Dr. Church will have written the foreword; his team has been very kind and compassionate so far and it's looking good. If not, please check back in a few weeks, I'll update with Version Two promptly. To check in on this manifestation, please email our support team at Support@ManifestingSecrets.com. I'll make sure you get a free copy of the new book with the foreword by Dr. Church.

Practicing Step Three

Today you're going to practice the "Healing Meditation" which I often refer to as the "drop" meditation. This is the absolute most simple, fundamental tool in your manifestation toolbox for relaxation and quantum manifestation.

First, lie down in a comfortable position; you may use a pillow under your head or legs. I love to meditate with a crystal blanket my friend Arthur Franklin gave me with Crystal Phi technology; it's very soothing.

Next, I want you to listen to the meditation in the bonuses section I've designed just for you at www.manifestingsecrets.com/bonuses

This is a long version of what I call the "Drop" meditation. I first began using a meditation like this after reading Dr. Kamini Desai's book called *Yoga Nidra* and practicing Yoga Nidra with her in her app of the same name. Then, Lauren Maloney Gepfert and the brilliant Susan Leety taught me how to use the drop meditation through their work helping people with paralysis or brain injuries heal through neuroplasticity.

I want you to commit this simple script to memory as your instant healing tool that you can use any time. Just 3-5 minutes of this meditation is an instant shift from your busy, tense, beta brain wave state to more relaxed, elevated vibrations and brain wave states conducive for healing and growing.

Remember: The Healing Meditation is an almost instantaneous shift. It doesn't involve complicated methodologies or complex systems. You don't need hours. You just need to drop.

Creating Intentions

"When you're connected to the power of intention, everywhere you go, and everyone you meet, is affected by you and the energy you radiate. As you become the power of intention, you'll see your dreams being fulfilled almost magically, and you'll see yourself creating huge ripples in the energy fields of others by your presence and nothing more."

WAYNE W. DYER

Manifestation is simply the act of changing the brain through mental intention. Mental intention is our fourth step in manifestation. Intention is what verbs we're putting to our thoughts.

The most fundamental intention is I AM. My favorite way to describe this comes from the late Dr. Wayne Dyer, known as the "father of motivation." He said that when we speak of ourselves, we speak of God. And so, therefore, would you say, "God is lazy, God is poor, God can't make money, God is lonely?" No. Rather, we would say, "God is rich. God is abundant. God is loving. God is giving. God is omnipotent."

The Bible says that all things are possible, and Jesus said, "let His mind be also in you." This is a direct reflection of how to think and live with intention. Let our thoughts be aligned with God's thoughts, or "Source," if you will.

We're born with innate abilities and gifts. Walking is a gift. Speaking is a gift. Peace of mind is a gift. Feeling and receiving love is a gift. As we grow up and become hurt, injured, traumatized, or paralyzed, we lose the superpowers most of us were born with.

However, we now know, thanks to techniques such as the functional MRI that our brain has the ability to change itself based on mental intention. We can actually see the brain light up and mental pathways rewire and remap through this scientifically sound technique.

Intention is not just a spiritual tool, but also a tool for neuroplasticity.

Our brains and bodies are truly miraculous. But as soon as you're hurt, your brain needs to relearn how to be in a healed state. This goes for healing emotional as well as physical wounds. Thanks to the process of neurogenesis, neurons grow every 21 days; and scientists can now trace this with functional MRIs. Fundamentally, this is neuroplasticity. Our brains can change themselves.

We can create new maps in our brains by using this gift of neurogenesis, or the new neurons that we have every 21 days, along with sensory input and mental intention.

When you combine sensory input and mental intention, you can heal your body and heart. The mentors who helped me grasp this concept are Neurofunctional Practitioners; they use neuroplasticity to heal spinal cord injury patients (those who are paralyzed) in swimming pools. If you read Norman Doidge's *The Brain That Heals Itself*, you will read dozens of studies of people who have healed themselves with this method of mental intention with sensory input.

One of the foremost reasons people become stuck in manifestation is because they focus on doing versus being. Allow me to explain. When you believe in yourself and your manifest power, you stop living in the energy of "doing." Your "I need," "I want," "I don't have" or "I would love to" become, "I like, I desire, I am, and so it is."

One of the reasons we fail to manifest is because we don't embody what we want to manifest. Remember: we don't attract what we want, we attract what we are. Instead of wanting or longing for something, become that thing in the here and now. When you say you want something, you are telling the universe you don't have it. You are admitting a sort of defeat. I want this, I lack this, I don't have this. When you live with the presence and faith that that thing has manifested, you attract that thing with your vibrations, receptivity, and especially with your gratitude.

It's so deeply satisfying to stop searching for what and/or whom you want and to start feeling the satisfaction of feeling and being what and whom your soul wants. I used to have this written on my mirror where I love to put my affirmations: "You are the one you desire. You are your soulmate. You have nothing to fear in this present now."

Sometimes the Law of Attraction oversimplifies manifestation by telling us that if we say or visualize something, that thing will appear. So much of the Law of Attraction is based on lack instead of embodying or being what we desire right now. Become what you want now, don't wait for a check in the mail or a romance to come along. Celebrate your faith, your being, your love, and your peace with the tools in this book. You have everything you need to be supremely happy. And after all, isn't happiness the ultimate goal?

Don't worry about not being perfect, but about remaining peaceful in your daily practice as you grow with kindness, patience, and encouragement. One of the best models I've seen to help us remember to Be instead of remaining in the lacking state of want is to remember this simple equation popularized by Zig Ziglar. Be: Do: Have.

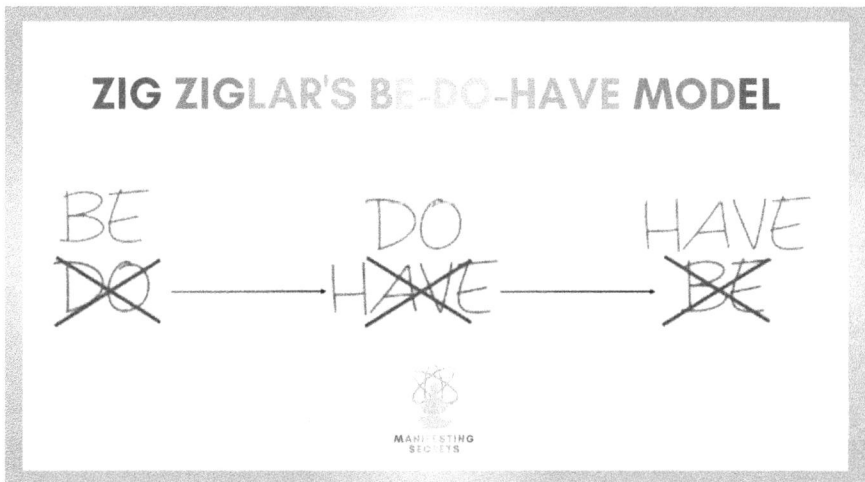

ZIG ZIGLAR'S BE-DO-HAVE MODEL

The reverse equation, Have, Do, Be, is the way we've learned how to operate in a negative world with performance pressures that values material versus spiritual wealth. For example: "When I have a million dollars, I will be able to do the things I want, such as visiting exotic places, and then I will be happy." Wrong. That's a Have, Do, Be equation that's contrary to manifestation.

"YOU'VE GOTTA BE BEFORE YOU CAN DO, AND YOU'VE GOTTA DO BEFORE YOU CAN HAVE."

ZIG ZIGLAR

However, the spiritual and manifesting equation with much more power is Be, Do, Have. First focus on BEING happy, BEING love, BEING abundant.

"Being" what you want precedes "getting" what you want. Our ego focuses on one thing: "what's in it for me." Our ego asks, "what can I get" versus "what can I give?"

However, true intention is propelled by feeling and being in the states of health, wealth, love, and happiness that we desire to manifest in the material world. That's right: true manifestation doesn't start with winning lottery tickets or fancy cars. Those material manifestations are generally preceded by embodying your desires and the feelings elicited by being what you want. And if you are not busy Being what and whom you desire, you'll likely piss away those "lottery wins" and wind up like most lottery winners... as depressed, impoverished, dissatisfied souls.

Your intention is already manifesting even before you state your intention or become consciously aware of it. The universe has already begun to provide what you desire without you having to speak a single word. Put another way, intention is your heart's desire. What you desire, you attract.

Practicing Step Four

Take a moment today to create intentions for what you can be.

Identify three Be, Do, Haves in your life.

"When I have $1,000,000, I will quit my job and I will be happy," is one example. Try to develop a few more unique ones.

Now flip the equation to state, "I AM _____." Commit to embodying the three states of being in your life today. At first it will seem awkward or even forced.

Remember, manifestation is a practice, not a magic pill. Keep practicing Being your peaceful, loving, radiant self and, in time, you will wire your mind to default to these wonderful states of Being no matter what is going on around you.

Breathwork for Manifestation

> *"Feelings come and go like clouds in a windy sky.*
> *Conscious breathing is my anchor."*
> THICH NHAT HANH

In this chapter, I honor Bel Carpenter. He has been my mentor in yoga, which includes asana or postures, meditation, and breathwork. He is the embodiment of a yogi and has taught me that yoga goes so far beyond the postures; yoga is a lifestyle. Bel often tells his students, "Practice yoga every day. Even if you don't come to class. Three to four hours is good, it's better than sleep." His students snicker at the suggestion that they "practice" yoga, just three four hours is good... But what Bel is illustrating is that breathwork, mindfulness, walking meditations, seated meditation, and asana are a three-to-four-hour practice spanned throughout his day. I have noticed that between my own asana practice, meditations while walking in nature, seated meditations, Yoga Nidra, and breathwork do, indeed, take place during several hours of my day more often than not. Now when everybody else snickers, I smile. Yoga is, in my opinion, the most profound tool on the planet for manifestation. Thank you, Bel, for being my Teacher.

Breathing is the fastest way to manifest a new reality. Perhaps that sounds a little crazy and hyperbolic, but in this chapter, I'll explain how.

Changing your entire life in any given moment can be as simple as taking a deep breath. At that moment, we tell our brains that everything is going to be okay. We move from reaction to response mode. Most importantly, taking a deep breath helps us to anchor into somatic awareness. Before responding with a subconscious program or "reaction" to the world around us, taking a deep breath helps us to see a greater truth, path, or strategy by bringing us out of the "mental" response zone and into awareness of what intuition is saying. Intuition is a body sensation, such as a ringing in the ear, a feeling in the "gut," a flutter of the heart, or even a throbbing in your big toe. Once we become aware of how these sensations feel in our bodies, we make better life choices based on the guidance our bodies provide.

Practice these breathing techniques first thing when you wake up and throughout your day as you make decisions. Take note of what your initial reaction might be to a situation and how you would traditionally respond or react to something. After taking a deep breath or spending just a few minutes in these breathwork exercises, notice how your responses and reactions to the world around you change.

1. Kapalbhati Breathing
[pronounced ka-pal-baa-tee]

"Kapal" means forehead and "bhati" means shining.

A long time ago, yogic breathing techniques were developed to help with well-being. These breaths fall under the category of "pranayama," or controlled breathing. In Sanskrit, pranayama means forehead shining breathing technique; or a breathing technique that gives you a shining intellect. "Prana" means breath or vital energy in the body; it's your life force. "Ayama" means control. The practice of controlled Kapalbhati breathing is a foundational breathing technique for health and vitality. What's more, Kapalbhati can feel very energizing.

How to Do It:

* To engage in Kapalbhati breath today, start in Sukhasana or the Easy Pose. It's what you may call "cross legged" in a seated position.

* Now focus your attention on your belly. Breathe in through your nostrils; this is a short, quick breath with your belly extending as you breath in.

* Breathe out with your nostrils as well. This breath will be short. Try 25 inhales and exhales for your first round.

* Sit and notice the sensations in your body with your mind still and quiet.

* Now try 3 additional rounds.

Kapalbhati helps your body generate heat, dissolve toxins, and improve kidney and liver function. It enhances blood circulation and digestion. It is also believed to help with your metabolism. Because Kapalbhati stimulates your abdominal organs, it helps give you a sense of balance and it is said to be helpful for those with diabetes.

Khapalbhati strengthens your lungs and is even believed to help cure asthma and sinus infections. Kapalbhati helps clear or activate chakras in your body (it can be very stimulating if you have stuck energy in your solar plexus or sacral chakras.) It is even said to help with anti-aging by removing stress or dark circles from your eyes.

* Be careful to stop if you feel dizzy and always talk to your physician if you have heart problems or a history of nausea or digestive disorders.

2) Alternate Nostril Breathing or "Nadi Shodhana"
[pronounced nah-dee-show-done-uh]

Nadi is a Sanskrit word that means "channel" or "flow." Shodhana means "purification."

Nadi Shodhana Pranayama is designed to purify the body and mind and restore a masculine/feminine balance. It almost instantly creates a harmony between the left and right hemispheres of your brain, helping you to get back into your "right mind" if you feel either too stressed or too lethargic or depressed.

How to Do It:

* Keeping mindful of your thumb and your middle finger, start by taking a deep breath with your fingers on your forehead.

* If you're using your right hand, you now breathe in your left nostril with your right nostril covered by your thumb. Use your middle finger to cover your left nostril and breathe out the right nostril.

* Now reverse; breathe in through your right nostril and out through your left; all the while keeping the nostril that you're breathing through open while the others closed by alternately your thumb or middle/ring finger.

Nadi Shodhana is very calming. As opposed to the short, quick breaths of Kapalbhati, Nadi Shodhana is performed more slowly. It is believed to improve concentration and mental clarity by providing equal amounts of oxygen to both sides of your brain. Try practicing Nadi Shodhana before an important phone call or meeting; or use it to calm down before bed.

3) Spinal Kriya
[pronounced cree-ya]

Imagine a line running from your tailbone to the top of your heads as you breathe. Traditionally from the Kundalini practice of Yoga, the Spinal Kriya can be performed a multitude of ways. This breathing technique is a psycho-physiological method that decarbonizes and recharges your blood with oxygen. The Spinal Kriya we shared is designed to bring a current of vitality to your brain and spinal centers.

How to Do It:

* Simply sit in Sukhasana or the Easy Pose. That's a cross-legged seat. Place your hands on your knees.

* Now focus on breathing in through your nose with a line of light directing from your tailbone or the base of your spine up to the tip of your forehead.

* Exhale while seeing that light move down from your forehead down your spine and back to the base of your spine.

4) Warrior or Ujjayi Breath
[pronounced ooh-jah-yee]

"Ujjayi" is a Sanskrit word which means "to conquer" or "to be victorious."

It is often translated to victorious or oceanic breath. The Warrior Breath or Ujjayi Breath is a method by which we breathe in and out slowly through our nose with awareness of a vibration or audible sound in the throat by constricting or tightening muscles in our throats. Your lips stay closed, but your exhale makes a "hah" sound.

This breath is intended to be long, smooth, and controlled. Because you can hear the breaths, it is helpful in bringing us to the present moment while building internal heat and warmth. Warrior Breath is a warming breath that helps to balance and stabilize you. It's used in the military to bring calming strength and presence of mind to participants.

This breath helps us increase the amount of oxygen to our bodies while relieving tension, helping to regulate blood pressure and detoxify our bodies. Lengthening exhale increases healing, and lengthening inhale builds power, renewal, and strength. Practice this breath before meditation, before phone calls for work, and even while driving in your car today.

How to Do It:

* Different from the previous breaths during which we inhaled short, quick breaths (Kapalbhati) and slow, even breaths through our nostrils (Nadi Shodhana), the Warrior Breath is a method by which we breath in quickly and out slowly through our nose.

* While the Kapalbhati breath generates heat in your body, the Ujjayi Breath cools you down.

Practicing Step Five

Today, practice these different breathing techniques while visualizing. Our visualization practice is going to be that of Recollection. Using a process of visualization will help you sink into rest and relaxation when you meditate.

First, try to identify a time during which you were thriving and loving life. It could be yesterday or even several years ago. Identify smells, sounds, habits, and health during that period.

Take time today to write down a list of 10 things that helped "propel" the best time of your life.

* Were you eating a high-frequency plant-based diet during that time?
* Performing swimming daily or rock climbing?
* What helped nourish the best version of you?
* Were you married? Single?
* Did you have children who were small? Grown?

Don't judge the things that contributed to that thriving life. Harness the feelings of those moments and capture that joy today.

We are magnets and our thoughts (and feelings) attract good or negative things to us. As you wrap up your visualization practice, use this simple mantra while you breathe. "I deserve the best, I accept the best, I choose the best."

While breathing deeply in a seated position with your palms resting on your knees facing upward, close your eyes and repeat your mantra for 3-5 minutes, twice today. See how you make more abundant decisions or feel more confident in your life as a result of this mantra.

Try putting your mantra in writing on your mirror, at your desk, or even on your fridge today to remind you that you deserve, accept, and choose the best.

STEP SIX:
Manifest Through Movement

"If you read ten thousand books but neglect to move your body in manifestation, you'll remain unchanged."

One of the reasons that Manifesting Secrets has graced the cover of Business Insider, been featured in Forbes, by Tony Robbins, in The Chicago Journal, New York Weekly, and so many other progressively minded articles and publications is because Manifesting Secrets harnesses the power of movement. There are programs on working out, programs on yoga, programs on dance, but Manifesting Secrets combines the science of movement with mind-rewiring like no other.

We understand the benefits of meditation and movement, but what about manifestation and movement? Movement is the critical key to unlocking our desires and creating permanent change in our bodies. When we move mindfully and with intention to heal our hearts, lives, and brain patterns, we supercharge our ability to manifest. That's because when you coordinate your brain with your heart or Spirit in movement, you engage each part of your brain, the visual, auditory, and kinesthetic (movement) executive functions.

Movement is critical to manifestation because relying on willpower alone will leave you unchanged. As much as you can read books or repeat mantras, movement is the key to producing physical, tangible results in the brain. Not only does intentional movement elevate your vibration, but it helps to identify and harmonize trauma, stuck energy, or pain from the body. Intentional movement shakes off emotional and psychological barriers that keep us stuck in subconscious programs that hurt our hearts, cause us to repeat failures or self-sabotage, and block us from confidently pursuing the lives we want to manifest.

Your body never lies. In the previous chapter we discussed how your body can be used as an antenna for truth through harnessing intuition. In addition, moving your body intentionally merges your mind, Spirit, and intention to ignite manifestation through helping the brain to fire and wire in a new direction. Movement serves to help us fully realize our emotions as we literally "shake off" our ego and surrender to the deep wisdom of our bodies.

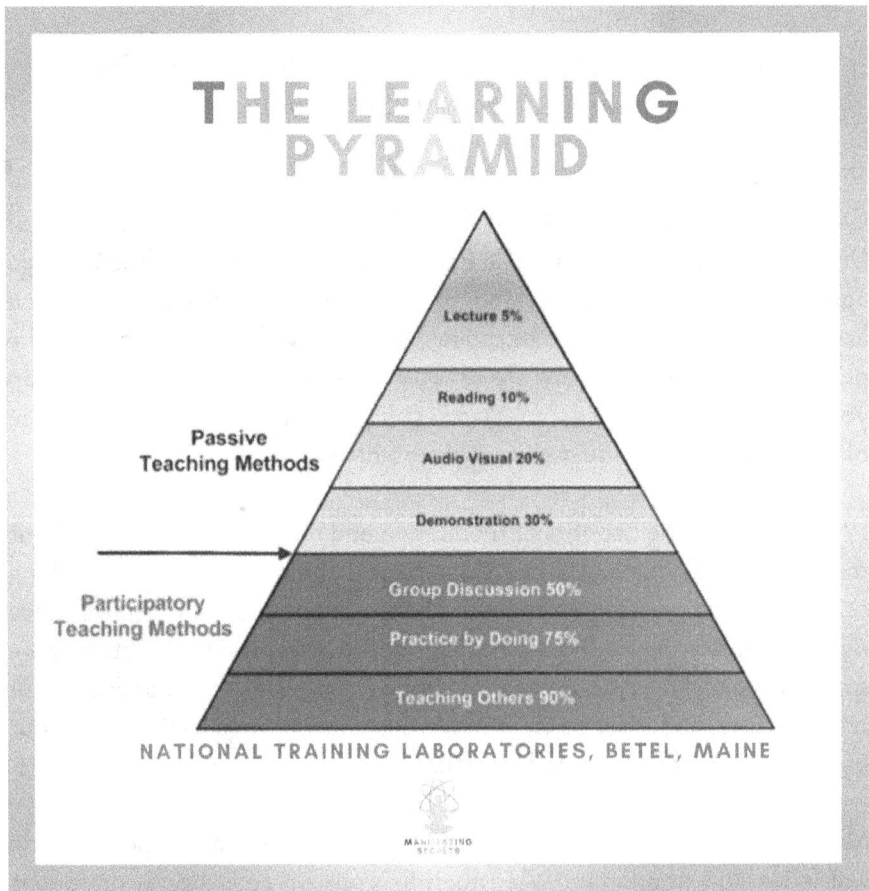

THE LEARNING PYRAMID

Passive Teaching Methods
- Lecture 5%
- Reading 10%
- Audio Visual 20%
- Demonstration 30%

Participatory Teaching Methods
- Group Discussion 50%
- Practice by Doing 75%
- Teaching Others 90%

NATIONAL TRAINING LABORATORIES, BETEL, MAINE

Above all, movement solidifies our manifestations by helping what we hear or intend to carve new pathways in the mind. Let's first look at how we learn. Did you know that when you read information, you only retain 10% of that information? However, the more you engage your senses, the more you retain. The National Training Laboratories in Betel, Maine suggest that when you do or teach information as opposed to merely reading it, you retain up to 90% of that information.

Simply put, when you engage more senses, you retain more. The Manifesting Secrets 90-Day Training incorporates several movement practices such as the Brain Integration Practice, light yoga practices, and chakra movement practices.

The reason for this isn't to develop a new group of yogis in the world, but to help you integrate the seal in the information provided in this course. Other ways to move with intention that you can start right now include free-form dancing or certain types of martial arts. The Neurofunctional Institute is known for using a technique called Sensory Stacking.

Sensory Stacking is the process of achieving what we want through mental intention combined with spoken word and movement. With sensory stacking, your brain can heal mental trauma, such as negative thought patterns that keep you stuck in bad habits, behaviors, or emotional distress. Your brain can also heal physical distress through combining intention with mindful movement. In fact, the founder of the Sensory Stack methods, Lauryn Gepfert, earned a degree in kinesiology and dance before she migrated her life's work into neuroscience and healing. She found that the two pair elegantly to create the most lasting, effective, and permanent change in the brain.

Experts believe that sealing your intention through movement helps turbo-charge your ability to manifest a new reality for your life and body. The theory states that when you visualize a result, say that result out loud, and then perform a movement to consecrate your intention, you're engaging more of the executive functions needed to achieve your desired manifestation and rewire neural pathways.

INTENTION =
SEE-SAY-DO

VISUALIZE THE RESULT.

VOICE THE RESULT.

ACHIEVE THE RESULT IN
YOUR BODY OR LIFE.

MANIFESTING
SECRETS

While moving your body, even through simple cross-body movements such as stroking your arms with opposite hands, you will find that your nervous system begins to calm down almost instantly. You will light up both sides of your brain and create a fertile environment for new brain patterns. You want to think about your intention, say your intention, and then move your body with intention is the formula for manifestation through movement.

CREATE A FERTILE ENVIRONMENT
FOR MANIFESTATION

VISUAL	AUDITORY	KINESTHETIC
MEMORY IMAGINATION VISUALIZATION MENTAL REHEARSAL PROPRIOCEPTION	INTERNAL VOICE (LISTEN TO THOUGHTS)	INTERNAL SENSORY 1. INTERPRET SENSATIONS 2. FEEL
EXTERNAL EYE 1. MOTOR NEURONS 2. SEEING	EXTERNAL VOICE (SELF-SPEAK)	EXTERNAL MOTOR DIRECTED FUNCTION

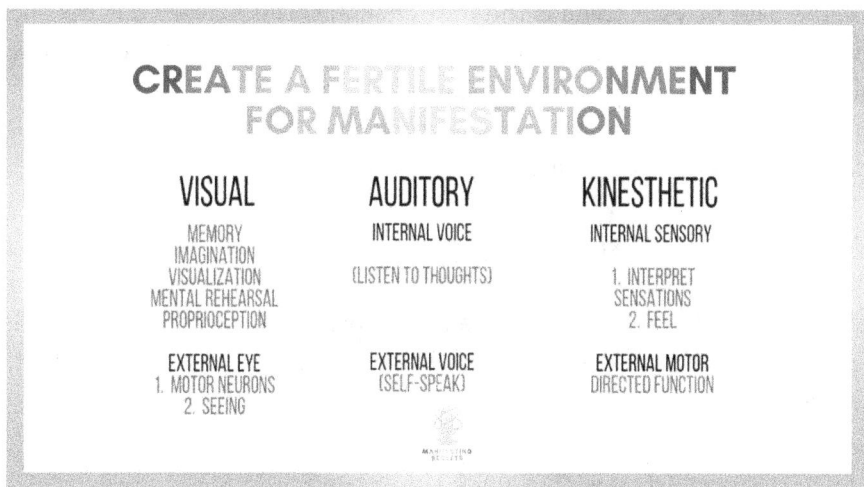

You may recall in Chapter Two that we discussed the power of neuroplasticity and defined it in two ways. Number one: cells that fire together wire together. Number two: what you don't use, you lose. Movement, by engaging the kinesthetic executive function of the brain, helps more neurons to fire... thereby helping more neurons to wire in the direction you choose.

Each time a group of neurons fire together and make a pattern, their tendency to fire again in the same pattern is increased because they pay closer attention to their associated neighbors. During times when I've experienced sadness, heartache, or anxiety, I have used this method of moving with intention to manifest a more positive outcome almost instantly.

Practicing Step Six

Today we'll practice just one of the seven Chakra Movement Practices from the Manifestation Card deck my Manifestation Secrets enjoy for daily manifestation practices.

At the base of your spine is the root chakra; this is where your sex and reproductive system exist; around your rectum or tailbone. The root chakra shows up in your body when financial needs, safety, and sexuality are off-balance. In order to harmonize your root chakra, practice grounding by dancing or even walking barefoot outside.

When your root chakra is balanced, you feel content, safe, stable, and peaceful. You will feel grounded and also free, like a dancer. An out of balance root chakra is typically seen in people who bounce from one task to another without self-care, which leads to burnout, feelings of defeat, lethargy, and even depression.

The word Muladhara is a combination of two Sanskrit words: "mula," which means "root" and "adhara," which means "support" or "base." Practice these postures while saying, "I am." If you visit www.manifestingcards.com, you'll find fifty-two practices to activate your brain's healing mechanisms.

Surrender

"Always say "yes" to the present moment. What could be more futile, more insane, than to create inner resistance to what already is? what could be more insane than to oppose life itself, which is now and always now? Surrender to what is. Say "yes" to life — and see how life suddenly starts working for you rather than against you."

ECKHART TOLLE

To manifest your best life, health, wealth and happiness, it is critical that you operate from a place of surrender, a combination of confidence and stillness. You may think of surrender as passive or even avoidant. Rather, surrender is an action, it is the powerful ability to let go.

Before we talk about the science of surrender, take a moment to read a portion of Dr. David Hawkins' book *Letting Go* that beautifully illustrates surrender:

"In this case, the desire was moderate and could, without much effort, be totally surrendered. By being totally surrendered, it was OK if the apartment happened, and OK if it didn't. Because of being totally surrendered, the impossible became possible; Manifesting Itself effortlessly and rapidly. We can all doubt this mechanism and look back at things that were achieved through effort, desire, achieving... and even obsessive, frenzied wanting."

The mind says, "Well, what if I had let go of the desire for those things? If it weren't for the desire, how would I have gotten them?" The truth is, we could have gotten them anyway. Only without anxiety and fear of not getting them.

Without all the energy expenditure. Without all the effort. Without all the trial and error. Without all the hard work. Well, the mind says if we got it effortlessly, how about the pride of achievement? Wouldn't we have to sacrifice that? Well, yes, we would have to relinquish the vanity of all that sacrifice and hard work that we put into it. We would have to give up the sentimentality about self-sacrifice and all the pain and hard work to achieve our goals.

This is a peculiar perversion in our society, isn't it? If we suddenly become successful almost effortlessly, then people are envious. It really annoys them that we didn't have to go through all kinds of anguish, pain, and suffering to get there. The mind believes that such anguish is the cost that must be paid for success.

Let's look at this belief: if it weren't for the negative programming that made us believe otherwise, why should we go through any cost of pain and suffering to achieve anything in our life? Isn't that a rather sadistic view of the world and the universe? Other blocks to the achievement of our wants and desires, of course, are conscious guilt and smallness. Peculiarly, the subconscious will only allow us to have what we think we deserve. The more we hang onto our negativity and the small self-image that results, the less we think we deserve.

And we unconsciously deny our souls the abundance which flows so easily to others. That is the reason for this saying, "The poor get poorer and the rich get richer." If we have a small view of ourselves, then what we deserve is poverty. And our unconscious will see to it that we have that actuality.

As we relinquish our smallness and re-validate our own inner innocence, and as we let go of resisting our generosity, openness, trust, loving us, and faith, then the unconscious will automatically start arranging life circumstances so that abundance begins to flow into our life.

You can read every book on the subject of manifestation, but if you don't take time to sit in silence with your own thoughts, you won't get anywhere fast. If you're familiar with the Bible, one of the Ten Commandments was to honor the Sabbath and keep it holy. Traditionally, this means to have a day of rest every week. In some traditional Jewish communities, families don't even drive cars on the Sabbath. This is a beautiful time because you best connect with your true self, desires, and strengths through rest.

I find that my best "work" is done when doing nothing at all. Meditation, silence, staring at my vision board, and spending time in nature doing walking meditations are when some of my best ideas and most peaceful moments occur. You see, if we don't take time every week (ideally for a period every morning right after we wake up) to simply "be," our minds will begin to succumb to mental chaos and incessant chatter. This incessant chaos and chatter often lead to depression or anxiety.

Do you ever feel like there's so much to do and you'll never get it all done? The truth is, you will never have more on your plate than you're capable of handling. When calibrating the 17 levels of consciousness on a scale of 1-1,000, Dr. David Hawkins found that the statement "There is no cause of anything" calibrates at one of the highest levels on earth; higher than anything else that could be conceivably calculated.

Practicing silence is a baby step toward a life of surrender. Silence can help us avoid reactions that bring unnecessary pain or harm to us and those around us. Practicing silence is a commitment to be content with simply being you; but also a practice of being content regardless of what others around you do.

When you meditate, take walks in nature, or simply sit still trying to remain consciously aware but unattached to the world around you, you will quiet the noise or turbulence in your mind. When you are first silent you may go a little crazy. Practice one or two hours a day with no television, no strategizing, no analyzing, no books or radio. Your mind will stop racing. Sure enough, the turbulence will quiet down. Your life will begin to surrender to you, the creator of your life.

The more you practice silence, the closer you are to the field of pure potentiality. You will become more aware of the world around you. And you will become less reactive to the chaos around you.

One of my favorite mantras to help me surrender in silence and stillness and listen for signs and synchronicities in my life is by Dr. Kamini Desai from her Yoga Nidra meditations.

It is simply, "I release my judgment about [name stressor] that causes stress, tension, and fear. I rest in calm awareness." Periodically, I will repeat this mantra for twenty-five whole minutes when I'm charged or triggered, which happens less and less on my course to becoming an evermore epic manifester. I often pair this with the mantra, "God is in control and He is working everything out in my favor."

When practicing the science of surrender, you can be calmly aware of your thoughts without attaching feelings of rage, anger, jealousy, loneliness, depression, victimhood, or even physical sickness that come up. Slow, gentle, silent breathing and meditation helps you rest with the feelings that come up and allows those thoughts and subsequent feelings to "just be." With the added calmness and awareness, you can begin to reverse engineer the thoughts that created those feelings.

For instance, perhaps you're feeling disappointed in yourself for overeating. This happens to me all the time when I crush massive bowls of popcorn for a nighttime snack. Stop for a moment and think about why you're overeating. Are you lonely? Are you feeling rejected? Do you lack confidence? Are you numbing out some pain that you're feeling today? Are you stuffing down your emotions with food? Are you in denial of pain or feelings you're having? Are you trying to hide, repress, or deny the truth of what you feel?

When you begin to more easily pinpoint the thoughts that create the feelings you have (and want to avoid), you can rewire your mind to create new thoughts and the life you desire. I like to take one day a month to "be silent." This is my version of a Sabbath, and when I haven't had one in a few weeks, I feel like I'm going to explode. On this silent day, I schedule no calls and only speak to my best friend that day; if at all. During this time, I don't "do" anything to propel my goals in life. I hike a lot, perhaps

I'll cook something nutritious and with a high vibe. Oftentimes, I'll do some gentle yoga and take a long nap with my coziest blankets and warmest incense. Periodically, I'll use this day for light fictional reading, an enema, painting my nails, a bath, hair conditioning treatments, and lots of music. I don't "work" on these days... and yet, my silence is the most important thing I can "do."

During these Sabbath days, I listen carefully to my heart and mind through my body or somatic awareness. I try to be very gentle with my body so that I can breathe deeply and feel everything it shows me all day. During these days, I find that I eat more intuitively and hear wisdom about different herbs, sprouts, vitamins, and minerals that my body wants me to feed it. I watch for downloads through signs, synchronicity, or God's audible voice during this time. As we learned in previous chapters, neuroplasticity experts have found through attaching participants to FMRI machines, that during silence and meditation, we learn more than any other time. Our brains light up like fireworks, and we even achieve the most neuroplasticity. That's right; we rewire our brains most effectively when we do absolutely nothing.

Did you ever notice your brain lights up right before bed? You aren't neurotic. On the contrary, you're experiencing the effect of neuroplasticity that comes from rest. During these acute moments of "awareness," I like to say mantras or meditate in order to direct my thoughts where I want them to go. This is a wonderful time to visualize how you want the following day to go or to rewire negative thoughts that arise.

I find that by resting in a place of feminine "surrender," I am more aware of and receptive to the miracles already in front of me. (You'll read in the upcoming chapters about feminine and masculine energies.) Again, it bears repeating: doing "less" gets more "done." Surrendering to the flow of life means accepting everything in it and allowing the good in the Universe to take precedence in your mind and life.

Surrendering is not passive; it's active.

Eckhart Tolle once said, "To some people surrender may have negative connotations, implying defeat, giving up, failing to rise to challenges, and so on. True surrender, however, is something entirely different. It does not mean to passively put up with whatever situation you find yourself in and to do nothing about it. Nor does it mean to cease making plans or initiating positive action. Surrender is the simple but profound wisdom of yielding to rather than opposing the flow of life."

Surrendering involves remaining open in our most feminine state. If you're a man, you still have both masculine and feminine characteristics in your actions and disposition. When we have an open posture of receptivity, which is a feminine characteristic, we attract good things. When we feel closed off, wary, suspicious, or have negative feelings, we attract those things from the world around us. Vibrate positive thoughts, act in positive ways, and you will create a positive life.

In September of 2020, I was hit with eight legal motions and contempt of court for innocuous and even false accusations made against me. Although the charges were absurd, the defense cost a tremendous amount of time and money. I lamented to friends and found myself playing victim for a moment. And then a friend said, "You have no choice but to surrender." By surrender, my friend didn't mean I needed to plead guilty to the charges. Rather, I created my case mindfully and calmly and then put my faith in God to handle things in the way He ought. I surrendered. I wrote on my mirror, "With God all things are possible" and "If God is for me, who can be against me?" I remembered each day that most critically to my case was to vibrate with love and receive the reflection of that love in the world around me. Above all, when I ceased to intertwine my own chaotic energy with the situation, answers became clearer to me. Surrender didn't cause me to give up, but rather to embrace my power.

One of my favorite books is *Seat of The Soul* by Gary Zukav. He does a great job of explaining how our surrender is not passive, but active. Now I understand that in order to truly manifest abundance, I must surrender. I must be like water. In my feminine flow, moving and receiving in faith. Less doing. Less masculine lists and activity. More receiving from a source with infinite resources and capabilities.

Gary writes:

"The 10% [of intention] that is choosing the path for the sake of health and wholeness has more power ultimately than the 90% that is fighting to remain where it is and have its own way. The Universe backs that 10% and not the 90%. ...As you move into the healing of who you are and the conscious journey toward what it is you want, recognize that the Universe backs the part of you that is of clearest intention. You are constantly receiving guidance and assistance from your guides and teachers, and from the Universe itself. When you choose consciously to move toward the energy of your soul, you invite that guidance. When you ask the Universe to bless you in your effort to align yourself with your soul, you open a passageway between yourself and your guides and teachers. You assist their efforts to assist you. You invoke the power of the non-physical world. That is what a blessing is."

We Want
to Hear from You!

Thank you for joining me on this journey of learning how to rewire your life with the power of your mind.

You are so powerful! The world is a better place because you have chosen to invest in yourself today.

Please head over to Amazon and leave us a quick review.

Your positive feedback will help me to keep writing and blessing the world with more books in the Manifesting series.

Head here now:
http://bit.ly/themanifestingmind

Closing

Powerful friend, thank you for joining me on the journey to learn about manifestation and the ways I use it in my life. Endless peace and happiness are available to those who use these tools mindfully. Don't stop. This is just your first step in creating a life of manifestation, and I invite you to choose one or all of the following ways to keep connecting, keep manifesting, and keep elevating your consciousness and the future of our world.

Join a Manifestation Mastermind to ask me
your most earnest manifestation questions live at
www.manifestingsecrets.com/apply

If you're ready to join me
in the Manifesting Secrets brain training,
please visit me now at
www.manifestingsecrets.com/

Download your FREE BONUSES
from this book in PDF and MP3 formats at
www.manifestingsecrets.com/bonuses

Follow me on Instagram at
www.instagram.com/stephaniepierucci/
www.instagram.com/momswearcapes
www.instagram.com/manifesting_secrets

Join the Brain Flow Yoga Tribe at
www.brainflowyoga.com

I'll be hanging out on Clubhouse at @spierucci

Don't hesitate to call our support team at
1-(855) 720-1111

If you prefer email, shoot me a line at
support@manifestingsecrets.com

On Facebook, join a community of supportive manifesters at
https://www.facebook.com/groups/manifestingsecrets/

Mind-Rewiring Affirmations

Please visit
www.manifestingsecrets.com/bonuses
to download your own Affirmations Posters

Words to Know

Affirmation
To "affirm" is to state that something is true. When applied to spiritual life, an affirmation is a statement of truth which one aspires to absorb into their life.

Alpha
Low frequency electromagnetic waves from 8 Hz to 13 Hz. They occur during relaxation, certain forms of meditation, and also after waking up and just before falling asleep. The ego is practically excluded at this stage, where extinction of consciousness is deeper and cause-effect relationships and logic are practically nonexistent.

Asana
The body postures and positions within the physical practice of Yoga used to achieve the goal of blending movement and postures with breath and uniting the mind and body and spirit.

Baader Meinhof Phenomenon
A phenomenon that occurs when a person, after having learned a fact, word, phrase, or other item for the first time, notices that they encounter that item again, usually several times, shortly after having learned it.

Beta
Low frequency electromagnetic waves from 12 Hz to 28 Hz. These are the waves of typical automated daily activity without any creativity or inventiveness.

Bioenergetics
A system of alternative psychotherapy based on the belief that emotional healing can be aided through resolution of bodily tension.

Confirmation Bias
The tendency to search for, interpret, favor, and recall information in a way that affirms one's prior beliefs or hypotheses.

Consciousness
At its simplest, it is a being's sentience or awareness of internal or external existence. It is one's connection with the universe.

Delta
Low frequency electromagnetic waves from 0.5 Hz to 3 Hz. Very deep sleep waves. The large areas of the brain are excluded.

Dis-ease
The hyphenated variant of disease used to place emphasis on the natural state of "ease" being imbalanced or disrupted.

Elevated Cognition
Elevated cognition is the process of choosing the thought that feels best.

Epigenetics
In biology, epigenetics is the study of heritable phenotype changes that do not involve alterations in the DNA sequence.

Functional MRI
A Magnetic Resonance Imaging procedure that measures brain activity by detecting associated changes in blood flow.

Gamma
Low frequency electromagnetic waves from 40 Hz to 100 Hz are waves of motion activity that are most often recorded during creative work and when we encounter a sudden daze when hit by inspiration; some believe that there are changes in how we perceive the world around us and our place in it.

Intention
The practice of bringing awareness and energy to a quality, virtue, or desire you would like to cultivate for yourself.

Love Frequency or 528 hertz
A frequency that is central to the "musical mathematical matrix of creation." In ancient and advanced traditions, the Love frequency is used to assist in manifesting miracles and produce blessings and is thought to resonate with DNA.

Manifestation
A means of creating or bringing something into being by repeatedly focusing on the right thoughts for your personal situation, staying in touch with your emotions, and changing any beliefs or habits that no longer serve you to be open to receive what you desire.

Mantra
In Hinduism and Buddhism, it is a word, words, or sounds repeated to aid concentration in meditation. A mantra is a sacred utterance, believed by practitioners to have psychological and/or spiritual powers.

Map of Consciousness
A map developed by Dr. David Hawkins using his Scale of Consciousness that is used for measuring positive from negative, power from force, and truth from falsehood. It is used to measure and effect change in quality of life and personal evolution.

Meditation
The process of quieting the mind for relaxation, regeneration, or religious/spiritual purposes. The goal is to attain an inner state of awareness and peace and intensify personal and spiritual growth.

Mindfulness
The psychological process of purposely bringing one's attention to experiences occurring in the present moment without judgment, which one can develop through the practice of meditation.

Mirror Neurons
Brain cells that react both when a particular action is performed and when it is only observed.

Mudra
Translates to "seal," "gesture," or "mark" in Sanskrit. They are symbolic gestures often practiced with the hands and fingers. Mudras facilitate the flow of energy in the subtle body, stimulating different parts of the physical body and mind, and are used to affect the flow of prana (life force).

Neurogenesis
The growth and development of nervous tissue when nervous system cells, the neurons, are produced by neural stem cells (NSC) s.

Neuroplasticity
The ability of the brain to form and reorganize synaptic connection and change continuously throughout an individual's life. Essentially, it is the brain's ability to "rewire."

Observer Effect
In psychology, a phenomenon where subjects alter their behavior when they are aware that an observer is present or that they are being recorded. Also known as the Hawthorne Effect.

Parasympathetic
Nervous system that functions to control the homeostasis of the body.

Pranayama
Prana means life force or breath sustaining the body. Ayama translates as "to extend or draw out." Put simply, pranayama is breath control.

Quantum Physics
The study of the behavior of matter and energy at the molecular, atomic, nuclear, and even smaller microscopic levels.

Reticular Activating System
A part of the mammalian brain located in the brain stem that in human biology is believed to play a role in many important functions, including sleeping and waking, behavioral motivation, breathing, and the beating of the heart.

Scale of Consciousness/ Map of Consciousness™
First illustrated by Dr. David Hawkins in his book *Power vs. Force*, showing that consciousness, and therefore each human emotion has a vibration in the exact same way that matter does. Heavy emotions like fear, anger or shame vibrate at low frequencies, while feelings of love, joy, and peace vibrate at high, uplifting frequencies.

Schumann's Waves
Electromagnetic waves generated by the planet's nucleus spread outward toward its surface and further toward the ionosphere, connecting all like an invisible thread to the matrix of the planet.

Self-Regulation
Behaviorally, self-regulation is the ability to act in your long-term best interest, consistent with your deepest values.

Self-Speak
The act or practice of speaking to oneself, either aloud or silently, consciously or unconsciously. A type of inner monologue in which you provide opinions, evaluations, or commentary on what you're doing or experiencing as it is happening.

Sensory Input
The response in a sensory organ when it receives stimuli.

Somatic
Somatics is a field within bodywork and movement studies which emphasizes internal physical perception and experience. The term is used in movement therapy to signify approaches based on the soma, or "the body as perceived from within."

Subconscious Emotional Program
The basis of healing through the power of your subconscious mind. The programs of your subconscious mind generate vibrational equivalency, and your body follows.

Sympathetic Nervous System
Nervous system that functions to mobilize the body's fight-or-flight response.

Theta

Low frequency electromagnetic waves from 4 Hz to 7 Hz. These are waves that occur during states of sleep, dreaming, hypnosis, meditation, trances, etc.

Yoga

Sanskrit word meaning yoke or union. A physical and mental practice that aspires to join the mind, body, and spirit using movement, breath, and awareness. The three parts of yoga are breathwork which connects you to the present moment. Asana or movement that helps to activate and release, heal, or reveal past traumas or pain, and meditation, a way to powerfully manifest your future. Traditionally defined, yoga is not merely the postures or asana, as Western culture may have had you believe.

Yoga Nidra

A Sanskrit term meaning "yogic sleep." Yoga Nidra is a conscious deep relaxation technique and form of meditation that provides the practitioner with intense physical and mental restoration.

Resources
for Further Study

A lot of years working with experts, scientists, coaches, therapists, and doctors has gone into the Manifesting Secrets program you see today.

In lieu of reading hundreds of books and spending tens of thousands of dollars on Manifestation Coaching like I did, we invite you to read one or all of these books that will help you increase your manifesting power in very specific areas.

Your challenge today is to obtain one of the books on the Manifesting Secrets reading list every month and share your thoughts in our Members Area on Facebook at https://www.facebook.com/groups/manifestingsecrets/

Books

33 Strategies of War
by Robert Greene

This book is an elegant illustration of how even the most cunning generals and military professionals engaged in a form of graceful ease under pressure. Through story, Robert Greene shows us how true leaders engage in deep breathwork or self-restraint to make the best decisions. If you're looking to live a life of more gentle restraint, ease, flow, and emotional regulation, this book is for you.

Activate Your Vagus Nerve
by Dr. Navaz Habib

Of the many books I've read on the Vagus Nerve, Dr. Habib's had the most information in the most concise way. He has included several fascinating and effective exercises to help you gauge the fortitude of your own Vagus Nerve.

Ask and It is Given
by Esther & Jerry Hicks

This is a decidedly wu-wu source or author with whom I still have a rather cautious relationship. Esther and Jerry Hicks channel wisdom and insight through a spirit named "Abraham" who has identified himself as a conglomeration of spirits. The insights that Esther and Jerry have channeled through Abraham range from wealth and prosperity coaching to relationship coaching. Abraham is a sort of ghost Tony Robbins in the best way. If you love the law of attraction and insights as thick and elegant as the Biblical Proverbs or a yoga guru, you'll love the work of Abraham-Hicks.

Atomic Habits
by James Clear

James Clear teaches us about building habits and how you can "eat the elephant" when you do it in small bites. This is a very encouraging piece for goal setting, entrepreneurship, and wealth creation, and it's full of great stories and anecdotes.

Attached.
by Dr. Amir Levine and Rachel S.F. Heller, M.A.

Learn about your attachment style and how you may thrive or sabotage your platonic and romantic relationships by falling into a habit that may not be serving yourself or your partner.

Becoming Supernatural
by Joe Dispenza

Dr. Dispenza is a Medical Doctor who healed a crippling back injury with the power of mental intention and meditation. Anything he's written to-date is strongly rooted in the power of meditation, mental intention, and filled with dozens of stories of how people like you and me can manifest through mental intention and neuroplasticity.

Bliss Brain
by Dr. Dawson Church

This is one of the most transformative books to understanding the neuroplastic power of meditation and the science of happiness. You will learn a method for meditation called "ego-meditation" as well as how to hack your brain for more bliss in your life, despite any circumstances.

Choose Yourself!
by James Altucher

Learn about how to begin pivoting old self-shaming or self-doubting programs and build small daily habits that lead to greatness. Pairs well with Atomic Habits and Days 1-3 of this program. The audible version of this book, read by Altucher, is fantastic and includes some bonus stories.

Code of the Extraordinary Mind
by Vishen Laksiani

This book is, aside from being a raw autobiography with engaging stories, a call to action for people who recognize that they are stuck in habits of scarcity and eager to step into their innate greatness.

Codependent No More
by Melody Beattie

Fundamental reading for relationships in today's world. It's the foundational book that defined a term that has become commonplace; Codependency. Codependency is a crippling psychological state of placating others before taking care of oneself. If you are in relationship trauma or exhaustion, please read this book.

Dodging Energy Vampires
by Dr. Christiane Northrup

We are all energy; relationships with friends, families or lovers who drain us are often what's holding us back from manifesting our dreams; and even keeping us sick. Learn how to identify and eradicate vampires in your world and watch your own energy levels soar.

Eastern Body Western Mind
by Anodea Judith

If you are interested in learning more about the chakras and how they have developed and play out in our everyday lives, this is a wildly comprehensive chakra analysis; said by many to be the most authoritative ever written.

Feminist Fight Club
by Jessica Bennett

For women looking to regain confidence in order to manifest the love, health, wealth and happiness they deserve, this book analyzes some of the language patterns and behaviors women use that hold them back; including body language (Day 27).

Flow
by Mihaly Csikszentmihalyi

Do you know what your flow state is? When you understand your flow state, you enjoy more happiness, satisfaction, and productivity with less expended energy or effort.

Getting Things Done
by David Allen

"Do it, delegate it, or delete it" is one of the principles Allen teaches in *Getting Things Done* that has impacted my own ability to surrender tasks or rituals that weren't serving my highest purpose. This is best for busy manifesters and businessmen and women looking to free up more space in their lives for abundance.

How to Have Confidence and Power In Dealing With People
by Leslie T. Giblin

Giblin identifies how to build relationships with others as they really are, not as you would like them to be.

Leadership & Self-Deception
by the Arbinger Institute

My friend Catherine Nomura actually told me to read this when I was having a relationship problem; although fundamentally it's a "business book," this was the first really profound book I read on personal responsibility. It serves to hold up a mirror to your shortcomings and their effects on others.

Lean In
by Sheryl Sandburg

Feminism "light;" Sandburg provides an encouraging and

decidedly politically correct book about women in the workplace; great for women seeking more confidence and men seeking more empathy for a woman's professional responsibilities post-family as well as the psychology that women carry that hinders them professionally.

Letting Go: The Pathway of Surrender
by Dr. David R. Hawkins
This is one of the most powerful books on surrender and non-attachment that you will ever read; it's a must-read.

Manifest Now
by Idil Ahmed
This book serves to provide some great law of attraction basics but doesn't dive into neuroscience very much. It serves as a sort of motivator and cheerleader to the journey; I pick it up when I want to stay focused on manifesting without thinking too hard.

Mating in Captivity
by Esther Perel
This book is one of the most profound illustrations of the real modern struggles in relationships, particularly as it pertains to infidelity.

Midnights with The Mystic
by Cheryl Simone and Sadguru Jaggi Vasudev
Delightful storytelling by Simone and the Audible version actually has Sadguru contributing to it and his laugh in the audio version is like sweet music to the soul. This book dives into fundamental philosophies of yoga, meditation, and even love.

Mindset: The Psychology of Success
by Carol Dweck
Why are some people successful and others are not? It's important that we as manifesters understand that mindset isn't just a tool, but it is THE tool to help us manifest our dream lives. This serves as a primer to neuroplasticity.

Mind to Matter
by Dawson Church
Outstanding introduction to neuroplasticity, a must-read if you're fascinated by rewiring the mind.

My Stroke of Insight
by Jill Bolte Taylor
Bolte Taylor had a stroke and near-miraculous healing, teaching herself to move and function again after being paralyzed after the incident. This is a profound story of healing.

Nonviolent Communication
by Dr. Marshall Rosenberg
In this book you'll learn how to manifest the outcomes you want in life with loving, compassionate and conscious communication.

Power Vs. Force
by David Hawkins
This book details the 20 years of muscle-testing studies he used to create his Scale of Consciousness.

Rethinking Narcissism
by Craig Malkin
Nothing kills the manifestation power of an empath like the common problem of becoming prey for narcissists. Enter: me. If you've struggled with narcissism or are in a relationship with a narcissist, you can develop compassion, understanding, and boundaries to heal with this book.

Seat of The Soul
by Gary Zukav
Gary Zukav was a leader in the New Age spirituality movement when Oprah Winfrey hosted him on her show in the 1990s. This is a fundamental book if you love New Age spirituality.

Seven Steps to Finding Your Life's Purpose
by Stephanie Pierucci
This is the Companion to Stephanie's training, "Seven Steps to Finding Your Life's Purpose." Please find it on Amazon here:

http://bit.ly/7purpose

Should I Stay or Should I Go?
by Ramani Durvasula, PhD
For individuals feeling stuck between a sense of duty to a partner and the sense that your relationship is not serving your highest self, this book will help you sort out your thoughts and weigh your options around living with and potentially leaving a narcissistic relationship.

Switch on Your Brain
by Dr. Caroline Leaf
Written by a pastor and neuroscientist, this book helps you understand how fundamental Biblical principles are designed to ignite your plastic brain and puts into perspective how Jesus' words were similar to modern-day neuroscientists.

The Big Leap
by Dr. Gay Hendricks
Dr. Hendricks discusses the concept of an "upper limit problem" and how many people hinder the flow of money, love, health, and

abundance into their lives because of their self-limiting beliefs.

The Biology of Belief
by Dr. Bruce H. Lipton
Dr. Lipton explains the biological functions behind manifestation. This is one of the most progressive books ever written on epigenetics and the concept of healing the brain to heal the body. It's positively foundational to the modern manifester.

The Brain That Changes Itself
by Norman Doidge, MD
This book is a foundational piece to understanding the emergence of neuroplastic healing in the modern day; it details stories of people who healed themselves with mental intention as well as the fundamental science behind neuroplasticity.

The Confidence Code
by Katy Kay and Claire Shipman
Confidence is critical to be able to manifest a new life with consciousness versus running on subconscious programs. Kay and Shipman question how much of confidence is based on genetics and how much is based on our own decision to be confident.

The Daily Stoic
by Ryan Holiday
This book is written like a 365-day devotional with short chapters from Stoic philosophers. It's paramount to my own discovery of emotional regulation and unlocking my ability to handle obstacles as opportunities.

The Emotional Life of Your Brain
by Richard J Davidson
Understand how to improve your emotional health as taught by a neuroscientist.

The Four Agreements
by Don Miguel Ruiz
This is basically my Ten Commandments: even though it's only four. I repeat one of these agreements every day, and it is one of the most important books I've ever read. I have made it my religion.

The Obstacle is The Way
by Ryan Holiday
It's almost impossible to walk away from Ryan Holiday's book *The Obstacle is The Way* without feeling like you can take on anything that comes your way. This isn't a mere survival guide for life, but it is the ultimate guide to making pain your power. The historical stories are fascinating and brilliantly told.

The Power of No
by James Altucher

Neuroscientists discuss the importance of strategically losing and strategically gaining in your mind-remapping practice. James discusses how important it is for an otherwise obsequious, people-pleasing society to learn how to protect themselves by saying No More.

The Power of Now
by Eckhart Tolle

This book is a fundamental reading in the power of being present in the moment and has been hailed as one of the most foundational philosophical works of the New Age.

The Power of Positive Thinking
by Norman Vincent Peale

If you've ever thought that positive thinking seemed like naiveté, Peale illustrates how to use this popular law of attraction concept to manifest in your life.

The Power of Vulnerability
by Brené Brown

We can't heal what we aren't being honest with. Learn how to love yourself, be honest with yourself and those around you, and draw healthy boundaries with this work by Brené.

The Science of Getting Rich
by Wallace Wattles

Learn from Wattles about how attracting money is an energetic practice of creating versus competition. He guarantees that getting rich is a science, not a lottery.

The Secret
by Rhonda Byrne

Although I pooh-pooh the simplicity of the Law of Attraction a lot, it's a wonderful tool to employ in your manifestation efforts. Use the Law of Attraction not as the entire engine to manifestation, but as the fuel that propels the emotions you'll need to manifest and rewire your mind. It's not the end-all be-all, but a critical supplement and piece of the puzzle.

The Spiritual Practice of Creating Income
by Susan Lustenberger

If you're experiencing scarcity, poor income, or money blocks, Susan discusses them from the spiritual perspective. One of the most impactful "ah-ha's" I've had about money comes from Susan's work. It truly changed my life and helped me to set boundaries that have elicited abundance. I send every entrepreneur friend this book and I'm perhaps its biggest fan.

The State of Affairs
by Esther Perel

If you're in a relationship that suffers a lack of faith or trust in one another, this book will help you understand the concept of sovereignty and give your partner the freedom they need to evolve and love you more deeply.

The Surrender Experiment
by Michael A. Singer

Surrender is a critical part of conscious manifestation. And it's hard to "learn" to "let go." Through stories of how he surrendered in order to build a billion-dollar company, Singer teaches the concept of letting go elegantly.

The Untethered Soul
by Michael A Singer

Learn how to free yourself and rise above your current boundaries and limitations with this book of ancient wisdom on inner peace.

The Truth
by Neil Strauss

Telling the truth isn't honored in a society that trains you to fit into a box. This book is a critical read for people looking for clarity in their own purpose as well as how to fit their own healing and sovereignty into a romantic relationship.

Think and Grow Rich
by Napoleon Hill

Foundational money mindset training: a business and wealth philosophy book that everybody must read.

Visioning
by John Assaraf

A great companion to creating a Vision Board or Mind Movie.

Wheels of Life
by Anodea Judith

Learn about the Chakras and other Eastern medicinal concepts that will help you to have more awareness around your body and energy centers. In the modern world, we interact with people and events with our heads, but what if we acted with more of our body's natural intuition and rhythms? This is a life of more inspiration, less anxiety, and more honesty. Pair this book with your studies from Day 29 of the Manifesting Secrets program.

Why She Buys
by Bridget Brennan

The psychology of why people purchase will help you expand your abundance and wealth.

Yoga Nidra
by Dr. Kamini Desai

We have so many beautiful yoga Nidras in the Manifesting Secrets course. This book helps you understand the power of restorative capabilities of this ancient yogic art of Enlightened Sleep.

You Are the Placebo
by Joe Dispenza
Written in 2014, this is the book that spearheaded a lot of the grassroots movement to heal through meditation and mental intention as Dr. Joe retells the story of how he healed himself from a paralyzing spinal cord injury. Another must-read if you love neuroplasticity.

Your Inner Physician and You
by John Upledger
If you haven't had the experience of cranio-sacral therapy, you're in for a treat. This is a gentle form of therapy that helps with soma-to-emotional release and optimal body-brain connection and has been attributed to many quite miraculous healings.

Lectures/ Webinars/ Talks

Alison Armstrong:
Understanding Women

Perhaps the most influential discussion I've heard on the primal and biological differences between men and women; why we do what we do and how we can more cohesively and easily partner with the opposite sex in life and love.

Amy Cuddy:
Your Body Language May Shape Who You Are

Use body language to help you rewire negativity in your mind that becomes more ingrained through body language. Your body language testifies your thoughts. Are you thinking thoughts of self-love, confidence, and safety, or do your thoughts wreak of insecurity, worthlessness, or imposter syndrome? My friend Blair Dunkley says that our behavior determines our thoughts, not the other way around. Use your postures and body language to manifest more confidence and positivity in your life. Every time you adjust your shoulders back or your chest forward... Every time you lift your head, speak more clearly, or engage in positive body language, try to accompany that with an affirmation. "I am powerful. I am capable. I am successful at everything I do. People want what I have."

The Discovery
by Dr. David Hawkins

As a longtime avid fan of Dr. Hawkins, I learned so much more about his spirit and who he was not just during his research years, but up to a few years before his passing. This book is a fundamental primer for anybody looking to increase consciousness and happiness in life.

Esther Perel:
Rethinking Infidelity

After beginning a relationship with a notorious "cheater" and then being cheated on by a man I absolutely adored, I began to study the concept of cheating and what it really means. Is it meant to hurt the other partner? Or do people cheat because they don't have the courage to say what they want or need? Is it because they are bored? Or because they love somebody enough to stay and lie? What is the law around cheating; or is every situation unique? A must-watch for anybody in a relationship.

Esther Perel:
The Secret To Desire in A Long term Relationship

Play with Polarity with this video from Esther Perel on keeping a man or woman "interested" and "turned on" by the relationship long after the chemical honeymoon period has ended.

Ram Dass:
The Original Be Here Now Talks

Ram Dass, born Richard Alpert, is the infamous psilocybin experimenter turned philosopher and speaker. He teaches you to be more present and less busy in your mind so that you can embody love, peace, and compassion.

Apps

NuCalm

NuCalm is a tool using symphonic music composed by neuroscientists to help heal physical trauma in the brain, help you get better sleep, and assist you with focus when you're awake. Visit NuCalm.com and use the code Stephanie for 10% off.

Yoga Nidra by Kamini Desai

Use this as a companion when learning about the power of meditation to release anger and heal your body through meditation.